T0135643

MPI Series in
Biological Cybernetics
No. 07, November 2003

MAX-PLANCK-GESELLSCHAFT

Titus R. Neumann

Biomimetic Spherical Vision

Biomimetische Algorithmen zum sphärischen Sehen

Bibliografische Information Der Deutschen Bibliothek

Die Deutsche Bibliothek verzeichnet diese Publikation in der Deutschen
Nationalbibliografie; detaillierte bibliografische Daten sind im Internet über
http://dnb.ddb.de abrufbar.

ISBN 3-8325-0410-9
ISBN 1618-3037

Logos Verlag Berlin
Comeniushof, Gubener Str. 47,
10243 Berlin
Tel.: +49 030 42 85 10 90
Fax: +49 030 42 85 10 92
INTERNET: http://www.logos-verlag.de

This work is dedicated to my grandfathers
  Heinrich Rau, Johann Mutterer, and Dr. Paul Neumann.

Acknowledgments

The work presented in this dissertation was done at the Max Planck Institute for Biological Cybernetics in Tübingen, Germany. It was funded by doctoral stipends from the Max Planck Society and from the Flughafen Frankfurt Main Stiftung. I wish to thank Prof. Heinrich H. Bülthoff for his encouragement and support, for his valuable criticism, and for providing an excellent research environment. I am also grateful to Prof. Andreas Zell from the Department of Computer Science at the University of Tübingen, for his helpful comments and for supporting my work. I am indebted to Prof. Hanspeter Mallot, Prof. Roland Hengstenberg, and my "doctoral grandfather", Prof. Karl Götz, for many stimulating discussions and their strong intellectual example. I owe many thanks to all the past and present members of the Bülthoff lab, especially to my early office mates Matthias Franz and Susanne Huber who introduced me to scientific life at the MPI. I thank Astros Chatziastros and Volker Franz for their support and for helpful discussions on data analysis and statistics; Björn Kreher for implementing the first version of my compound eye model; Tobias Neumann for his valuable feedback on many aspects of my work; and Kathrin Schrick for critical reading. I thank Markus von der Heyde, Bernhard Riecke, Daniel Berger, Douglas Cunningham, Ian Thornton, Wolfgang Hübner, Wolfgang Stürzl and Kerstin Stockmeier for stimulating and interesting conversations, and also Mario Kleiner and Arnulf Graf for theoretical discussions after midnight. Many thanks to Isabelle Bülthoff for organizing the lab's supply of opera tickets, and to all the members of our glorious "Mad Max" 100 km relay running team, including Stefan Braun, Volker Franz, Barbara Knappmeyer, Bernhard Riecke, Sibylle Steck, and Alexa Ruppertsberg. I am grateful to Walter Heinz, Michael Renner and Philipp Georg who kept my computers running, and especially to Dagmar Maier for her organizational skills and invaluable help. Finally I thank my parents, my brother, and my wife Kathrin for their love and support throughout my studies.

VI

Zusammenfassung

Biomimetisches Sehen (*biomimetic vision*) ist ein neuer, interdisziplinärer Ansatz im maschinellen Sehen. Er beruht auf der Beobachtung, dass viele natürliche Organismen im Lauf der Evolution hochoptimierte visuelle Strategien zur Kontrolle ihres Verhaltens entwickelt haben. Diese Strategien sind bestehenden technischen Lösungen weit überlegen und können in einer neuen Generation künstlicher Systeme dazu verwendet werden, eine Vielzahl von Kontrollproblemen zu lösen. In dieser Dissertation beschreibe ich ein vollständiges, biomimetisches Sehsystem mit sphärischem Blickfeld. Mehrere Algorithmen und Datenstrukturen zur effizienten Erfassung, Repräsentation, Verarbeitung und verhaltensorientierten Auswertung sphärischer Bilder werden vorgestellt. Die Algorithmen haben die verschiedenen Verarbeitungsstufen im visuellen System fliegender Insekten zum Vorbild und werden im Kontext einer visuellen Flugstabilisierung getestet.

Die Komplexaugen fliegender Insekten sind hochoptimiert für die Aufnahme flugrelevanter Informationen aus der Umgebung. Zu den typischen Abtast- und Filtereigenschaften dieser Augen gehören ein sphärisches Sichtfeld, diskrete Rezeptoreinheiten, eine geringe räumliche Auflösung, überlappende, gaußförmige rezeptive Felder, sowie eine ortsvariante Verteilung der lokalen Blickrichtungen. In dieser Arbeit stelle ich ein exaktes und effizientes Simulationsmodell für die Optik von Insektenaugen mit singulärem Projektionszentrum vor. Es erzeugt diskrete, sphärische Bilder aus mehreren ebenen, perspektivischen Ansichten der Umgebung. Die Sensitivitätsverteilungen der rezeptiven Felder werden auf diese ebenen Ansichten projiziert, um perspektivische Verzerrungen auszugleichen. Im Gegensatz zu anderen, auf Raytracing basierenden Simulationen für Insektenaugen verwendet dieser Algorithmus Hardware-beschleunigte Computergrafik und erzielt daher auch in komplexen und realistischen virtuellen Umgebungen hohe Bildwiederholraten.

Das visuelle System belegt einen großen Teil des Gehirns von Insekten. Seine regelmäßige, retinotope Struktur erhält lokale Nachbarschaftsbeziehungen im sphärischen Sehfeld über mehrere aufeinanderfolgende Schichten aufrecht. Dies deutet auf einen massiv parallelen, vorwärtsgerichteten Informationsfluss hin, an dem hauptsächlich lokale Interaktionen wie örtliche Mittelung, laterale Inhibition und lokale Korrelation beteiligt sind. Zu den Aufgaben der retinotopen Schichten gehören lokale Bildverarbeitungsoperationen wie raumzeitliche Kontrastverstärkung und lokale Bewegungsdetektion. Um den prinzipiellen Aufbau des visuellen Systems von Insekten auf das maschinelle Sehen zu übertragen, stelle ich neue Datenstrukturen und Algorithmen für die effiziente und homogene Repräsentation und Verarbeitung sphärischer Bilder vor. Ein sphärisches Bild wird als Menge diskreter Datenpunkte definiert, die sich auf den Knoten eines ikosaeder-basierten geodesischen Gitters befinden, wodurch ein Raster hexagonaler Bildelemente entsteht. Richtungsbezogene Information wie der lokale optische Fluss wird durch die Kanten des Gitters repräsentiert. Ausgehend von dieser graph-basierten Repräsentation beschreibe ich lokale Bildverarbeitungsoperationen für räumliche und zeitliche Tiefpass- und Hochpassfilter, Auflösungspyramiden sowie lokale Bewegungsdetektion für diskrete sphärische Bilder.

Das grundlegende Stabilisierungsverhalten fliegender Insekten umfasst unter anderem Lagekontrolle, Kursstabilisierung, Kollisionsvermeidung und Höhenregelung. Die meisten dieser Verhaltensmuster sind stereotyp und nahezu rein reaktiv, jedoch extrem schnell. Dies deutet auf einen vorwärtsgerichteten Informationsfluss mit nur wenigen sequentiellen Verarbeitungsschritten zwischen sensorischem Eingang und Motoraktivierung hin. Ich verwende diese Architektur, um in einem künstlichen autonomen Agenten mit biomimetischem sphärischem Sehsystem verschiedene insekteninspirierte, visuelle Orientierungsstrategien zu implementieren. Das zentrale Verarbeitungselement in jedem Kontrollmechanismus ist eine von Tangentialneuronen in Fliegen inspirierte Großfeld-Integrationseinheit. Diese Einheiten haben komplexe rezeptive Felder, die Vorwissen über invariante Eigenschaften der Umgebung und des Verhaltens enthalten. Sie reagieren hochselektiv auf verhaltensrelevante Muster in der sphärischen Verteilung der von vorausgehenden, retinotopen Verarbeitungsstufen empfangenen, lokalen Eingangssignale. Ich zeige, wie eine optimale Sensitivitätsverteilung direkt aus der Korrelation der lokalen Eingangssignale mit einem bestimmten Verhalten hergeleitet werden kann, und präsentiere optimale rezeptive Felder für die Schätzung der räumlichen Lage, der Rotation um die Vertikalachse, der relativen Nähe von Hindernissen, sowie der relativen Nähe der Geländeoberfläche. Ich zeige, dass diese rezeptiven Felder genügend Informationen für eine dreidimensionale, visuelle Flugregelung mit allen sechs Freiheitsgraden einschließlich Hindernisvermeidung und Geländefolgeverhalten in einer hochrealistischen, virtuellen Umgebung bereitstellen.

Die in dieser Dissertation vorgestellten, insekten-inspirierten Algorithmen zur Gewinnung und Verarbeitung sphärischer Bilder eignen sich für eine Vielzahl innovativer Anwendungen. In biologischen Studien ermöglichen sie die Implementierung von exakten und effizienten Simulationsmodellen des visuellen Systems von Insekten, wodurch ein leistungsfähiges Werkzeug für die Analyse und Rekonstruktion von visuell gesteuertem Verhalten bereitgestellt wird. Im maschinellen Sehen stellen biomimetische Verarbeitungs- und Kontrollstrategien einfache, aber hocheffiziente Alternativen zu herkömmlichen Ansätzen dar, und könnten für zukünftige Technologien wie der autonomen Robotik im Miniatur-Bereich sowie in der Luft- und Raumfahrt von entscheidender Bedeutung sein.

Abstract

Biomimetic vision is a novel, interdisciplinary approach to machine vision. It is based on the observation that many natural organisms have evolved highly optimized visual strategies to control their behavior. These strategies are far superior to existing technical solutions, and can be used in a new generation of artificial systems to solve various control tasks. In this dissertation I present a complete, biomimetic vision system with a spherical field of view. I introduce various algorithms and data structures for the efficient acquisition, representation, processing and behavior-oriented evaluation of spherical images. The algorithms are inspired by the different processing stages in the visual system of flying insects, and are tested in the context of visual flight stabilization.

The compound eyes of flying insects are highly optimized for the acquisition of flight-relevant information from the environment. Typical sampling and filtering properties include a spherical field of view, discrete receptor units, low image resolution, overlapping Gaussian-shaped receptive fields, and a space-variant distribution of the local viewing directions. In this study I present an accurate and efficient simulation model for singular-viewpoint insect eye optics which generates discrete spherical images from multiple planar, perspective views of the environment. The sensitivity distributions of the receptive fields are projected onto these planar views to compensate for perspective distortions. In contrast to other insect eye simulations based on ray tracing, this algorithm employs hardware accelerated computer graphics and can therefore be used in complex and realistic virtual environments at high frame rates.

The visual system occupies a large part of the insect brain. Its regular, retinotopic structure preserves local neighborhood relations in the spherical field of view throughout several successive layers. This indicates a massively parallel, feed-forward flow of information, involving mainly local interactions such as local averaging, lateral inhibition and local correlation. The retinotopic layers are devoted to local image processing operations such as spatiotemporal contrast enhancement and local motion detection. In order to transfer the design principles of the insect visual system to machine vision I introduce novel data structures and algorithms for the efficient and homogeneous representation and processing of spherical images. A spherical image is defined as a set of discrete data points located on the vertices of an icosahedral geodesic grid, resulting in a raster of hexagonal pixels. Directional information such as local optic flow is represented by the edges of the grid. Based on this graph representation I describe local image processing operations for spatial and temporal low pass and high pass filtering, resolution pyramids and local motion detection for discrete spherical images.

Basic stabilization behaviors of flying insects include attitude control, course stabilization, collision avoidance, and altitude control. Most of these behaviors are stereotyped and almost purely reactive, but extremely fast. This indicates a feed-forward flow of information with few sequential processing steps between sensory input and

X

motor activation. I use this type of architecture to implement various insect-inspired, visual orientation strategies in an artificial autonomous system with biomimetic spherical vision. The central processing element of each control mechanism is a wide-field integration unit inspired by fly tangential neurons. These units have complex receptive fields containing prior knowledge on invariant properties of both the environment and the behavior. They respond in a highly selective manner to behaviorally relevant patterns in the spherical distribution of local input signals received from preceding, retinotopic processing stages. I show how an optimal sensitivity distribution can be determined directly from the correlation of the local input signals with a specific behavior, and present optimal receptive fields for the estimation of attitude, rotation about the vertical axis, relative nearness of obstacles, and relative nearness of the terrain surface. I demonstrate that these receptive fields provide sufficient information for three-dimensional, visual flight control with all six degrees of freedom, including obstacle avoidance and terrain following in a highly realistic virtual environment.

The insect-inspired algorithms for spherical image acquisition and processing proposed in this dissertation are well-suited for a multitude of innovative applications. In biological studies, they allow the implementation of accurate and efficient simulation models of the insect visual system, thereby providing a powerful tool for the analysis and reconstruction of visually controlled behavior. In machine vision, biomimetic processing and control strategies provide simple but highly efficient alternatives to traditional approaches, and may be crucial for future emerging technologies such as autonomous aerospace and miniature robotics.

Contents

Chapter 1

Introduction

1.1 The Biomimetic Approach

Biomimetic[1] systems are artificial structures designed for a specific purpose by imitating aspects of natural organisms. The imitation may occur on various levels including morphology, processing structures, and control strategies. It is based on the observation that natural organisms have evolved highly optimized solutions for a multitude of engineering problems. Thus, biological findings may provide crucial ideas for the design of artificial systems. Conversely, artificial systems may be used in biological studies to model aspects of natural organisms. By definition, building biomimetic systems is an interdisciplinary approach. It requires background knowledge of both the biological example and the target application. In this dissertation, I apply the biomimetic principle to the problem of spherical vision, i.e., the acquisition, processing, and behavior-oriented evaluation of spherical images. All proposed algorithms and data structures are inspired by the visual system of flying insects, and are designed in the context of visual flight control. As an interdisciplinary study, this dissertation includes concepts from biology, physics, machine vision, robotics and computer graphics.

1.2 Spherical Vision

Long before the age of machine vision and autonomous robots, J. J. Gibson employed spherical vision to describe the fundamental operating rules for visually controlled spatial behavior in animals and humans (Gibson, 1950, 1958). The central element of his theory of locomotion is the closed loop of action and perception: A stimulus not only affects the control of behavior, but behavior in turn controls which stimulus is perceived. Thus, an animal can generate a particular behavior by eliciting a specific

---

[1]*mimetic:* from Greek μίμησις, *imitation, copy*

1

stimulus pattern, and inhibit other behaviors or events by avoiding the corresponding patterns in the stimulus.

According to Gibson, all visual stimuli an animal can possibly perceive are contained in the flux of light which fills the environment after being reflected many times and in all directions by the surfaces of objects. All rays converging from arbitrary directions to a specific point in the environment constitute the *optic array*. An eye located at that particular point can register the *static pattern* of the optic array which is formed by the different light intensities or spectral frequencies received from different directions. The static pattern is unique for each location in the environment and changes continuously during locomotion to a different location. The changes in the static pattern form the *flow pattern* of the optic array. The flow pattern is unique for a specific path of locomotion, but independent of the specific static pattern that carries it. Both the static pattern and the flow pattern are stimuli for the visual control of locomotion relative to the objects in the environment (Gibson, 1958).

In modern, mathematical terms, Gibson's "flux of light filling the environment" is defined by the *plenoptic function*

$$I = P(\xi, \psi, \lambda, x, y, z, t) \tag{1.1}$$

(Adelson and Bergen, 1991; McMillan and Bishop, 1995). This function describes all radiant energy I within a band of spectral wavelengths λ which can be perceived from the current viewpoint (x, y, z) at time t. The viewing direction is defined by the azimuth angle $\xi \in [-180, 180]$ and the elevation angle $\psi \in [-90, 90]$. The "static pattern of the optic array" is a complete sample of the plenoptic function at a specific location in the environment and constitutes a full *spherical image*. The "flow pattern of the optic array" can be mathematically described as a *spherical optic flow field*. In particular, self-motion-induced optic flow is determined by the current translatory and rotatory motion of the observer, and by the distance to the object perceived in the specific viewing direction (Koenderink, 1986; Koenderink and van Doorn, 1987).

Compared to eyes or cameras with a limited field of view, full spherical vision has a number of advantages. In a spherical field of view, all objects in the environment are simultaneously visible as long as they are not occluded or too far away to be resolved (Gibson, 1958). This facilitates the orientation relative to stationary objects such as landmarks, and allows the detection and tracking of moving objects such as predators or prey anywhere in the surrounding environment without eye or head movements. Further, spherical optic flow fields always contain the regions of smallest and largest absolute image motion. For translatory self-motion of the observer, the optic flow is smallest at the foci of expansion and contraction. For rotatory self-motion, the smallest flow occurs at the poles of rotation. In both cases, the optic flow is largest on the great circles perpendicular to the axes of translation or rotation. The simultaneous visibility of all regions facilitates the task of identifying these axes. Moreover, spherical optic flow fields induced by translatory and rotatory self-motion are always distinct (Nelson and Aloimonos, 1988; Dahmen, Wüst, and Zeil, 1997). Translatory flow has the same direction on opposite sides of the sphere, whereas rotatory flow has opposite directions on opposite sides. In a limited field of view covering less than one hemisphere the distinction between rotatory and translatory flow is generally more difficult.

In essence, a full spherical field of view captures the maximum amount of visual information available at a particular point in the environment, and is especially useful for the control of three-dimensional spatial behavior such as visual flight stabilization and navigation.

1.3 Learning from Insects

Spherical vision plays a central role among the various sensory modalities controlling the spatial behavior of flying insects. Despite the low resolution of their compound eyes, ranging from 700 receptor units per eye in the fruit fly *Drosophila melanogaster*, to 28000 in the dragonfly *Anax junius* (Land, 1997b), flying insects exhibit a multitude of visually controlled, highly sophisticated three-dimensional flight maneuvers such as stabilizing flight against external disturbances (Götz, 1968; Reichardt, 1969; Collett and Land, 1975a; Collett, 1980b; Collett, Nalbach, and Wagner, 1993), chasing fast flying prey or mating partners (Srinivasan and Bernard, 1977; Collett, 1980a; Bülthoff, Poggio, and Wehrhahn, 1980; Wehrhahn, Poggio, and Bülthoff, 1982; Kirschfeld, 1997), avoiding obstacles during flight (Tammero and Dickinson, 2002; Kelber and Zeil, 1997), hovering on the spot (Zeil and Wittmann, 1989; Kelber and Zeil, 1990), and landing on objects and surfaces (Borst and Bahde, 1988; Srinivasan, Zhang, Chahl, Barth, and Venkatesh, 2000). In addition, some insects are capable of visual homing and navigation (Collett and Land, 1975b; Cartwright and Collett, 1983; Srinivasan and Zhang, 2000). Considering the small size - less than a cubic millimeter - and low energy consumption of their brains, insects easily outperform any existing artificial system in each of these tasks.

The following properties of the insect visual system are assumed to contribute to this superior performance:

1. Insect compound eyes have an approximately spherical field of view. As described in the previous section, this allows to capture the maximum amount of visual information available at a specific location in the environment.

2. The insect visual system is optimized for motion vision rather than shape perception (Land, 1997b). The low image resolution is sufficient for the computation of local optic flow while minimizing the required amount of local image process-

ing. The spherical distribution of optic flow is a rich source of flight-relevant information (Gibson, 1950, 1958).

3. The early layers of the insect visual system exhibit a regular, retinotopic structure preserving local neighborhood relations in the spherical field of view throughout several successive layers. The predominant tasks of these layers are spatiotemporal signal enhancement and local motion detection at every location in the spherical field of view (Laughlin, 1989). This indicates that the insect brain performs image processing operations such as local averaging, lateral inhibition and local correlation, in a massively parallel fashion.

4. Wide-field integration units such as the lobula plate tangential neurons of the fly (Krapp and Hengstenberg, 1996; Krapp, Hengstenberg, and Hengstenberg, 1998) allow to recover flight-relevant information in a highly selective manner directly from the spherical distribution of local intensity or motion signals. These units have complex receptive fields containing prior knowledge on invariant properties of both the environment and the behavior (Franz and Krapp, 2000). In addition, wide-field integration makes the insect visual system robust against noise and failures of single light receptors or processing units.

5. Many basic stabilization behaviors of flying insects such as the dorsal light response for attitude control (Hengstenberg, Sandemann, and Hengstenberg, 1986; Schuppe and Hengstenberg, 1993) or the optomotor response for course stabilization (Götz, 1968; Reichardt, 1969; Hausen, 1993; Hausen and Egelhaaf, 1989) are stereotyped and almost purely reactive or reflex-like. This indicates feed-forward information processing without complex internal representations.

6. The extremely fast behavioral responses of insects to visual stimuli indicate that only few sequential processing steps are involved between image acquisition and motor activation. The direct neuronal chain connecting the eyes with the flight muscles consists of only six to seven successive cells (Hausen and Egelhaaf, 1989), allowing extremely short delay times between stimulus perception and reaction.

7. Flies are mechanically tolerant against possible failures of the visual system. In contrast to most robot systems, a fly can survive multiple collisions with walls or windows without damage.

In short, numerous anatomical, electrophysiological and behavioral studies indicate that the entire visual system of flying insects is highly specialized on the extraction of flight-relevant information from the environment. Applying the processing and control strategies found in flying insects to artificial systems may lead to a new generation of biomimetic, autonomous systems with major improvements in performance and robustness.

1.4 Dissertation Objective

In this dissertation I introduce novel, insect-inspired algorithms and data structures for the efficient acquisition, processing, and behavior-oriented evaluation of discrete spherical images in the context of visual flight control. In biological studies they allow to implement accurate and efficient simulation models of the insect visual system, and thereby provide a powerful tool for the analysis and reconstruction of visually controlled behavior. In machine vision, the proposed processing and control strategies provide simple but highly efficient alternatives to traditional approaches, and have various potential applications including autonomous vehicle guidance, driver or pilot assistance devices, as well as aerospace and miniature robotics.

1.5 Overview

This dissertation is organized according to the sequence of processing stages in the visual system of insects. Each chapter focuses on a specific aspect of spherical vision, and begins with information on the biological background, then introduces insect-inspired computational models and implementations, and finally discusses results, performance and applications. Chapter 2 describes an accurate and efficient simulation model for insect compound eyes in complex and photorealistic virtual environments. The eyes of flying insects are highly optimized for the acquisition of flight-relevant information from the environment, and their specific sampling and filtering properties constitute the initial processing stage in the insect visual system. Chapter 3 introduces novel data structures and algorithms for the efficient and homogeneous representation and processing of discrete spherical images. They closely resemble the topological and functional organization of the early, retinotopic layers in the insect visual system, and allow various local operations such as spatial and temporal low pass and high pass filtering by lateral interaction, as well as local motion detection. Wide-field integration units inspired by fly tangential neurons are presented in Chapter 4. These units extract behaviorally relevant information from the spherical distribution of local intensity and motion signals. I show how optimal receptive fields for attitude control, course stabilization, collision avoidance, and altitude control can be determined directly from the correlation of the local input signals with the corresponding behavior, and demonstrate that these apparently simple, visual orientation strategies are fully sufficient for three-dimensional, autonomous flight control and terrain following. The dissertation concludes with a brief summary and gives perspectives for future extensions and applications. Appendix A discusses the acquisition of spherical images by tetrahedral camera arrangements in computer simulations and real-world implementations.

Chapter 2

Modeling Insect Compound Eyes - Space-Variant Omnidirectional Vision

The compound eyes of flying insects are highly optimized for the acquisition of flight-relevant information such as three-dimensional self-motion and orientation, as well as the spatial layout of the environment. Thus, artificial vision systems confronted with comparable tasks may benefit from imitating their design principles. In this chapter I present an accurate and efficient simulation model for compound eye optics which generates low resolution spherical images from multiple planar, perspective views of the environment. The sensitivity distributions of the receptive fields are projected onto these planar views to compensate for perspective distortions. In contrast to other insect eye simulations based on ray tracing, this algorithm includes hardware accelerated computer graphics and can be used for complex and realistic virtual environments at high frame rates. Applications of this approach can be envisioned for both the reconstruction of visual stimuli seen by natural insects and for the development of novel, biomimetic vision strategies in artificial systems.

2.1   Introduction

The compound eyes of flying insects are the primary source of information for a variety of visual orientation strategies. Behaviors such as flight stabilization, obstacle avoidance, object fixation, detection of prey and predators, tracking of mating partners, as well as visual homing and navigation are based on visual control mechanisms. Many of these mechanisms are amazingly simple but highly robust and efficient, since the entire visual system of insects is optimized by natural evolution to selectively acquire and process only such information which is relevant for the visual control of specific behaviors and specific lifestyles in typical environments. Since compound eyes constitute the initial processing stage of the insect visual system, they play an important role

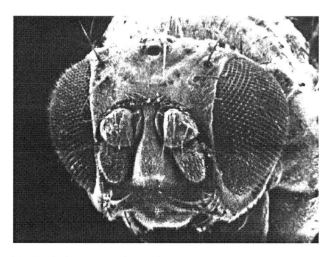

Figure 2.1: Head of male *Drosophila melanogaster*. (Reprinted, with permission, from M. Heisenberg and R. Wolf: *Vision in Drosophila*, p. 11, Fig. 1. © 1984 Springer-Verlag.)

in the selection of perceived stimulus properties and in the pre-filtering of acquired information for subsequent processing steps. Artificial systems may strongly benefit from imitating the insect visual system, including not only the neural mechanisms and control strategies, but also the design principles of insect compound eyes. Thus, an accurate and efficient computational model for compound eyes is a prerequisite for the development of novel, insect-inspired computer vision systems specialized on tasks like visual flight control and navigation. In addition, a detailed and accurate simulation model for compound eye optics provides a powerful tool for biological studies, allowing comprehensive computer simulations of insect vision and behavior. In open-loop studies, the stimulus seen by a specific insect in a particular experiment can be accurately reconstructed and used for further analysis, whereas closed-loop simulations can be used to test models of visually controlled insect behavior.

2.1.1   Insect Compound Eyes

Insect compound eyes fascinate many researchers because they are very different from our own, human eyes (Fig. 2.1). Since compound eyes are rather small objects, detailed examinations could only begin with the development of the microscope. It is reported that in 1610 Galileo used his telescope in a reverse manner to observe insect eyes at

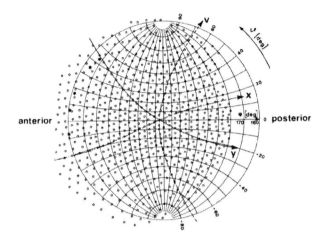

Figure 2.2: Distribution of optical axes in the left eye of *Drosophila melanogaster*. Each circle indicates the viewing direction of a single ommatidium. (Reprinted, with permission, from M. Heisenberg and R. Wolf: *Vision in Drosophila*, p. 11, Fig. 2. © 1984 Springer-Verlag.)

close range (Burkhardt, 1989). Presumably the first detailed drawings of insect heads and compound eyes were published 55 years later in Robert Hooke's "Micrographia", which showed that insect eyes are composed of a multitude of distinct facets (Hooke, 1665). In subsequent anatomical and histological studies the term *ommatidia* was introduced for these facets. An important step in understanding their exact function was the mosaic theory by Johannes Müller (1826): A single ommatidium is not a complete miniature lens eye, but measures only the intensity of the light entering parallel to its own direction. There is no further spatial resolution within a single ommatidium. Thus, each facet corresponds to one tile in the mosaic of the resulting image, and the overall resolution of the eye is determined by the number of ommatidia.

With his pioneering work on the dioptrics of compound eyes, Sigmund Exner (1891) founded the field of physiological optics and provided the basis of modern compound eye research. He distinguished two basic types of compound eyes: Apposition eyes directly implement the mosaic theory since each receptor unit has its own lens, whereas in superposition eyes parallel light entering through multiple neighboring corneal lenses and crystalline cones is focused on each receptor unit. Another type of compound eyes, the neural superposition eye of dipteran flies, has an open rhabdom containing multiple receptor cells with separate rhabdomeres (Kirschfeld, 1967). Light entering a particular ommatidium through its corneal lens is directed to the cen-

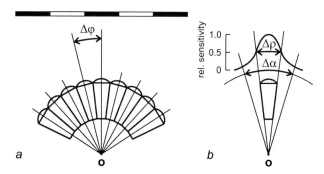

Figure 2.3: Visual acuity of compound eyes. *a.* The interommatidial angle $\Delta\varphi$ defines the spatial sampling frequency. *b.* For each receptor unit, the half-width of the Gaussian-shaped sensitivity distribution is defined by the acceptance angle $\Delta\rho$ (Götz, 1965; Land, 1997). In the simulation model presented here the receptive field size is restricted by an aperture angle $\Delta\alpha$.

tral receptor of the same ommatidium only if the direction of incidence coincides with its optical axis. If the direction of incidence is parallel to the optical axis of one of the adjacent ommatidia, the light is directed to a receptor in that particular ommatidium. Thus, each receptor of an ommatidium shares the same viewing direction as one of the receptors in each of the six neighboring ommatidia. Since the axons of these receptors converge, the resulting neural image is the same as in other types of compound eyes.

Mallock (1894) and Barlow (1952) emphasize the important role of diffraction as a limiting factor for the visual acuity of compound eyes. Consequently, the spatial acuity of a single ommatidium can only be improved by increasing its lens diameter. Thus, a higher overall resolution would not only require a larger number of facets, but also an increased diameter of each single ommatidium, which together would result in an impracticable eye and head size. Insects circumvent this problem by limiting the high spatial resolution to small 'acute zones' while the remaining regions of the eye have a much coarser resolution, resulting in a space-variant receptor distribution.

More details on the history of compound eye research are given by Goldsmith (1989) and Burkhardt (1989). The different types of compound eyes are comprehensively reviewed by Nilsson (1989). Land (1989) focuses on the variations in the structure and design of compound eyes, and van Hateren (1989) describes the lens and photoreceptor optics of single ommatidia.

At present, insect compound eyes are anatomically and physiologically well understood (Land, 1997a). Their fundamental sampling and filtering properties are illustrated in Fig. 2.3 and can be summarized as follows:

- *Discrete receptor units.* Compound eyes are composed of multiple discrete facets (ommatidia) arranged on a regular grid. Each ommatidium represents one 'pixel' of the perceived image, receiving light from a specific viewing direction through one (apposition eye) or multiple lenses (superposition eye). The local viewing directions of adjacent receptor units are separated by the interommatidial angle $\Delta\varphi$ (Götz, 1965; Horridge, 1992).

- *Low image resolution.* Compared to human lens eyes which have a foveal acuity of approximately $\Delta\varphi = 1$ arc minute, the spatial resolution of compound eyes is low. The approximate number of ommatidia ranges from 700 in *Drosophila* with $\Delta\varphi = 5°$, and 3000 in *Calliphora* with $\Delta\varphi = 1.3°$, to a maximum of 28000 in the dragonfly *Anax junius* with $\Delta\varphi = 0.24°$ (Land, 1997b).

- *Spherical field of view.* In flying insects the field of view covers the entire surrounding environment, eventually with exception of a small, caudal region obscured by the body. For some species, e.g., the fruit fly *Drosophila melanogaster* (Buchner, 1971), the water strider *Gerris lacustris* (Dahmen, 1991), or the blowfly *Calliphora erythrocephala* (Petrowitz, Dahmen, Egelhaaf, and Krapp, 2000), accurate maps of the receptor distribution on the sphere have been determined. The distribution of optical axes in Drosophila is shown in Fig. 2.2.

- *Singular viewpoint.* In the context of flight control it can be assumed that the viewing directions of all receptor units of both eyes originate from a singular nodal point, since an eye and head diameter of 1-3 mm is very small compared to typical object distances in the magnitude of decimeters or meters. Thus, only few insects with a large interocular distance have the neural circuitry for stereo vision, and even in those cases stereo range finding is not possible for distances beyond a few centimeters (Schwind, 1989).

- *Spatial low pass filtering.* Spatial wavelengths shorter than $2\Delta\varphi$ cannot be resolved unambiguously. To avoid aliasing effects, high spatial frequencies are suppressed using overlapping, Gaussian-shaped receptive fields with the acceptance angle $\Delta\rho$. Although full anti-aliasing is only achieved for a ratio of $\Delta\rho/\Delta\varphi \geq 2$, in many insects this ratio is $\Delta\rho/\Delta\varphi \approx 1$ to increase contrast (Götz, 1965; Land, 1997b; Horridge, 1992).

- *Space-variant receptor distribution.* The spatial acuity of compound eyes is limited by light diffraction. A higher resolution requires both an increased number of ommatidia and larger lens diameters. Since an overall increased resolution would lead to an impracticable eye and head diameter many compound eyes have local 'acute zones' in which the resolution is enhanced up to five times compared to other regions of the eye (Land, 1997b).

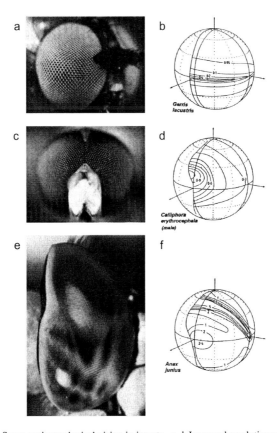

Figure 2.4: Space-variant spherical vision in insects. *a, b* Increased resolution at the horizon in the empid fly (*a*) and the water strider *Gerris lacustris* (*b*). *c, d* Frontal or frontodorsal acute zones in the male hoverfly *Syritta pipiens* (*c*) and the male blowfly *Calliphora erythrocephala* (*d*). *e, f* High overall resolution and double acute zones in the dragonflies *Aeshna multicolor* (*e*) and *Anax junius* (*f*). (*a* Photograph courtesy of Dr. Jochen Zeil. *c* Reprinted, with permission, from M.F. Land: Variations in the structure and design of compound eyes. In D.G. Stavenga and R.C. Hardie (Eds.): *Facets of Vision*, p. 102, Fig. 9. ⓒ 1989 Springer-Verlag. *e* Reprinted, with permission, from T.E. Sherk: Development of the compound eyes of dragonflies (Odonata). III. Adult compound eyes. *Journal of Experimental Zoology*, Vol. 203. ⓒ 1978 John Wiley & Sons, Inc. *b, d, f* Reprinted, with permission, from M.F. Land: Visual acuity in insects. *Annual Review of Entomology*, Vol. 42. ⓒ 1997 Annual Reviews.)

- *Ecological and behavioral specialization.* The size, shape and location of the acute zones is highly optimized for typical behaviors in specific environments (Fig. 2.4). For instance, insects living in flat environments or flying closely above the water surface, such as desert ants (Zollikofer, Wehner, and Fukushi, 1995), water striders (Dahmen, 1991) or empid flies (Zeil, Nalbach, and Nalbach, 1989), have an increased resolution in a narrow band around the horizon. In contrast, most male flies as well as predatory flies are equipped with pronounced frontal acute zones to track females and prey. Some dragonflies have dual acute zones in the frontal and in the frontodorsal regions (Sherk, 1978). The latter is assumed to be involved in prey detection.

In summary, insect compound eyes provide low resolution, space-variant, spherical vision with a singular viewpoint. They do not constitute a general purpose vision system, but are adapted to specific behaviors in typical environments and provide a highly optimized information selection and pre-filtering for subsequent processing stages. Experimental results and observations from insect ethology indicate that the visual information acquired through compound eyes is fully sufficient for a highly effective flight control and navigation.

2.1.2 Artificial Insect Vision: What does a fly see?

Land (1997a) describes what it would mean to see the world with the eyes of a bee. Since these eyes have a resolution about 100 times coarser than our own, one would have to stretch out an arm and imagine the world divided into pixels the size of the little fingernail. A similar description was already proposed 100 years earlier by Mallock (1894): "The best of the eyes... would give a picture about as good as if executed in rather coarse wool-work and viewed at a distance of a foot." (cited after Land (1997b)). The first attempt to technically reconstruct an insect's view of its environment was made by von Uexküll and Brock (1927) who presented a photograph reproduced with an increasingly coarser dot raster to give an impression of the low spatial resolution of compound eyes (Fig. 2.5). The broad availability of high performance computer technology in recent years provides an entirely novel and powerful tool for compound eye research. The automatic evaluation of computational models of insect vision and behavior has become feasible, including simultaneous, dynamic simulations of complex visual stimuli and eye models.

A number of computational models for compound eye optics have been developed during the past decade. Cliff (1991) showed a first, ray tracing-based computer simulation of an idealized hoverfly compound eye looking at a simple black and white stripe pattern. Giger (1995) demonstrated what static planar images might look like for a honeybee. The images were projected on a virtual eye surface and downsampled by the

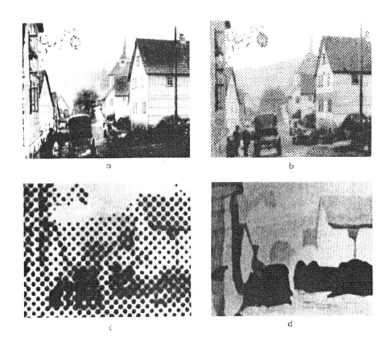

Figure 2.5: View of an old village. *a* Photography. *b*, *c* The same view reproduced with increasingly coarser resolution. *d* Scheme of what the view might look like to an insect. (von Uexküll and Brock, 1927)

ommatidial grid. A study by Vorobyev, Gumbert, Kunze, Giurfa, and Menzel (1997) applied the spectral properties and the hexagonal raster of insect eyes to planar images of plants and flowers. van Hateren (2001) investigated the spatiotemporal processing in fly photoreceptors and visual interneurons. He calculated neural images occurring in an array of simulated large monopolar cells from a sequence of planar and fish-eye-distorted input images recorded in a simple virtual environment. In a recent simulation study, Tammero and Dickinson (2002) approximate the visual acuity of *Drosophila* by generating a Mercator projection of the surrounding environment and applying a regular sampling grid with $5°$ spacing along azimuth and elevation to the resulting image, averaging over patches of $5° \times 5°$. The common goal of these studies is an accurate reconstruction of particular stimulus properties as seen through a specific insect eye.

The resulting images are used in an open-loop fashion for visualization (Cliff, 1991; Giger, 1995) or further analysis and environmental image statistics (Vorobyev et al., 1997; van Hateren, 2001; Tammero and Dickinson, 2002).

Aspects of compound eye optics are also implemented in various insect-inspired computer vision systems designed for the autonomous control of spatial behavior in a closed control loop. Examples are studies on an optic flow-based altitude control using a one-dimensional receptor array with a space-variant distribution of local viewing directions (Mura and Franceschini, 1994), on two- and three-dimensional corridor following and obstacle avoidance based on a minimal number of receptors (Huber, Mallot, and Bülthoff, 1996; Neumann, Huber, and Bülthoff, 1997; Dale and Collett, 2001), on view-based navigation using one-dimensional, low resolution panoramic images (Franz, Schölkopf, Mallot, and Bülthoff, 1998a), and on three-dimensional flight control evaluating the local optic flow in selected regions of a spherical field of view (Neumann and Bülthoff, 2001).

All of these eye models are restricted by one or more of the following constraints:

- The field of view is limited and does not comprise the entire sphere (Giger, 1995; Vorobyev et al., 1997; van Hateren, 2001; Mura and Franceschini, 1994; Franz et al., 1998a; Neumann and Bülthoff, 2001).

- Point sampling or insufficient filtering may lead to spatial aliasing (Cliff, 1991; Giger, 1995; Mura and Franceschini, 1994).

- Regular sampling grids in the Mercator plane result in systematic errors due to polar singularities (Tammero and Dickinson, 2002). As a consequence, the image resolution is lowest around the horizon and increases towards the dorsal and ventral poles. Furthermore, the largest EMD[1] base angles as well as the largest solid angles per receptor unit occur at the horizon, the smallest at the poles. Thus, a regular sampling grid in the Mercator plane introduces severe distortions in the spherical image. However, the error produced by this sampling strategy remains small if the visual stimulus is restricted to regions around the horizon.

- Since most of the above applications are based on ray tracing, both the eye model and the visual stimulus are strongly simplified to speed up processing. The efficient computation of the visual input is crucial in particular for the closed-loop evaluation of large populations, e.g., in studies involving genetic algorithms (Huber et al., 1996; Neumann et al., 1997; Dale and Collett, 2001).

- Ray tracing is inefficient and slow for complex scenes and large numbers of samples since it is not supported by current computer graphics hardware.

[1]elementary motion detector

In the following I present a novel computer simulation model for insect compound eyes that is both more accurate and more efficient than previous approaches. The model allows to specify arbitrary spherical distributions of receptor viewing directions and filtering properties, and is capable of reconstructing an insect's view in highly realistic virtual environments at high frame rates.

2.2   Compound Eye Model

The compound eyes of most flying insects have an approximately singular viewpoint and a spherical field of view. Their primary function is to map all regions of the environment visible from the current eye position onto a spherical retinal image. Thus, the involved sampling and filtering operations have to be defined on the sphere.

2.2.1   Space-Variant Image Processing on the Sphere

Both the currently visible environment and the retinal image have a common center of projection and can be represented on the finite, two-dimensional surface of a sphere, parameterized by the azimuth angle $\xi \in [-180, 180]$ and the elevation angle $\psi \in [-90, 90]$. A local viewing direction originating from the center of the sphere is denoted as a pair

$$(\xi, \psi) \in [-180, 180] \times [-90, 90] \subset \Re^2 \,. \tag{2.1}$$

However, this notation bears a number of problems such as polar singularities and ambiguities for $\psi = \pm 90°$, or the difficult determination of the angle between two distinct local viewing directions. These problems do not occur if each viewing direction is described by a three-dimensional unit vector

$$\mathbf{d} \in U = \{\mathbf{u} \in \Re^3 \,|\, \mathbf{u} \cdot \mathbf{u} = 1\} \tag{2.2}$$

which can be determined from an azimuth-elevation pair using

$$\mathbf{d}(\xi, \psi) = \begin{pmatrix} \cos \xi \cos \psi \\ \sin \xi \cos \psi \\ -\sin \psi \end{pmatrix} \,. \tag{2.3}$$

Note that both notations parameterize a two-dimensional spherical surface. The angle

$$\angle(\mathbf{d}_1, \mathbf{d}_2) : U^2 \to [0, 180] \subset \Re \tag{2.4}$$

between two local viewing directions $\mathbf{d}_1$ and $\mathbf{d}_2$ is defined as

$$\angle(\mathbf{d}_1, \mathbf{d}_2) := \arccos(\mathbf{d}_1 \cdot \mathbf{d}_2) \,. \tag{2.5}$$

Extending a notation for space-variant neural mapping (Mallot, von Seelen, and Giannakopoulos, 1990) to the spherical domain, the set of all possible viewing directions originating from the current eye position in the environment is denoted as source area S, the set of all directions covered by the retinal image as target area T, with

$$S, T \subseteq U \,. \tag{2.6}$$

The current source and target images are defined by the functions

$$I_S(\mathbf{d}) : S \to \Re \quad \text{and} \quad I_T(\mathbf{d}) : T \to \Re \,, \tag{2.7}$$

respectively, assigning a gray value to each local viewing direction on the sphere. An extension for color images is straightforward, using either local vectors of multiple color components instead of scalar gray values, or separate spherical images for each color channel.

The compound eye model determines the retinal image from the surrounding environment by a space-variant linear operator

$$I_S(\mathbf{d}_S) \longmapsto I_T(\mathbf{d}_T) := \int_S I_S(\mathbf{d}_S)\, K(\mathbf{d}_S; \mathbf{d}_T)\, d\mathbf{d}_S \tag{2.8}$$

integrating over the two-dimensional, spherical source domain S. For each local viewing direction $\mathbf{d}_S \in S$ in the source area and each direction $\mathbf{d}_T \in T$ in the retinal target area the space-variant kernel

$$K(\mathbf{d}_S; \mathbf{d}_T) : S \times T \to \Re \tag{2.9}$$

specifies the weight by which the input stimulus influences the retinal image.

A kernel closely approximating the spatial low pass filtering properties of compound eye optics is the Gaussian function (Land, 1997b). Here, an approximation of an isotropic Gaussian kernel is defined on the sphere as a function of the angular distance $\angle(\mathbf{d}_S, \mathbf{d}_T)$ between a specific source and target viewing direction. The kernel

$$K(\mathbf{d}_S; \mathbf{d}_T) := c(\mathbf{d}_T) \cdot K'(\mathbf{d}_S; \mathbf{d}_T) \tag{2.10}$$

is composed of the space-variant normalization factor

$$c(\mathbf{d}_T) := 1 / \int_S K'(\mathbf{d}_S; \mathbf{d}_T)\, d\mathbf{d}_S \tag{2.11}$$

ensuring a uniform gain for all receptive fields, and the truncated Gaussian-shaped function

$$K'(\mathbf{d}_S; \mathbf{d}_T) := \begin{cases} e^{-\frac{2}{\Delta\rho^2(\mathbf{d}_T)} \angle^2(\mathbf{d}_S, \mathbf{d}_T)}, & \angle(\mathbf{d}_S, \mathbf{d}_T) < \frac{1}{2}\Delta\alpha(\mathbf{d}_T) \\ 0, & \text{else} \end{cases} \tag{2.12}$$

with local acceptance angle $\Delta\rho(\mathbf{d}) : T \to \Re$ and local cutoff angle $\Delta\alpha(\mathbf{d}) : T \to \Re$. Both angles are space-variant and depend on the local viewing direction $\mathbf{d}$.

In the following I show how this general model can be adapted for efficient computer simulations of insect compound eyes in highly realistic virtual environments.

Figure 2.6: Example scene on a cube environment map. The environment map is composed of six square perspective images, shown unfolded (*upper right*) and on a cube surface (*lower left*).

2.2.2   Omnidirectional World Projections

Current 3D computer graphics technology is capable of generating photorealistic views of complex scenes at high frame rates. Thus, it provides a powerful tool for the evaluation and test of visual processing algorithms and eye models. It is based on hardware-accelerated, polygon-based raster graphics optimized to produce perspective views in planar, rectangular images composed of discrete pixels. However, the compound eye model described above requires spherical input images which cannot be generated directly by this technology. In this section I describe an algorithm that solves this problem by representing the surrounding environment as a combination of multiple planar perspective images entirely enclosing the eye point.

The most intuitive arrangement is the cube environment map (Greene, 1986) composed of six square perspective views of the environment, one for each face of a cube centered at the origin of the eye coordinate system (Fig. 2.6 and 2.7). All six images need to be updated for all changes of the eye point location or of the environment. As an alternative, the tetrahedron environment map is discussed in Appendix A. It comprises only four perspective images, which is the minimal number of planes required to enclose a point in 3D space. The algorithm described in this section can be used for both configurations.

The gray or color value in a specific viewing direction is determined by intersecting the viewing direction with the corresponding face of the environment map. To avoid aliasing effects due to point sampling, Greene and Heckbert (1986) proposed the elliptical weighted average filter, a Gaussian-shaped, concentric weight distribution around the intersection point in the image plane. However, this approximation deviates from the correct, spherical weight distribution which is defined as a function of the angular distance from the viewing direction (Fig. 2.8). Since the error increases with the angular width of the filter mask, the elliptical weighted average filter is not suitable for the large ommatidial acceptance angles of compound eyes. In the following the correct spherical filtering transformation is described.

2.2.3 Discrete-Pixel Filtering Transformation

Both the environment map and the compound eye retinal image are composed of discrete pixels, referred to as source pixels $p \in P_S = \{1, 2, \ldots, |P_S|\}$ and target pixels $q \in P_T = \{1, 2, \ldots, |P_T|\}$, respectively. $|P_S|$ is the total number of pixels in the environment map, $|P_T|$ the number of receptor units in the eye model. The source image is represented by the one-dimensional vector $I_S \in \Re^{|P_S|}$ containing the entire environment map. Each source pixel p has a gray value $I_S[p]$. The target image $I_T \in \Re^{|P_T|}$ contains one gray value $I_T[q]$ for each receptor unit q.

The receptive field of receptor unit $q \in P_T$ is described by the weight vector

$$\mathbf{w}_q^{\mathrm{Rcpt}} = \left(w_{q,1}, \ldots, w_{q,p}, \ldots, w_{q,|P_S|} \right)^{\mathrm{T}} \tag{2.13}$$

indicating the contribution of each input pixel value $I_S[p]$ to the output value $I_T[q]$. The weight matrix $\mathbf{W}^{\mathrm{Rcpt}} \in \Re^{|P_T| \times |P_S|}$ is defined as

$$\mathbf{W}^{\mathrm{Rcpt}} = \left(\mathbf{w}_1^{\mathrm{Rcpt}}, \ldots, \mathbf{w}_q^{\mathrm{Rcpt}}, \ldots, \mathbf{w}_{|P_T|}^{\mathrm{Rcpt}} \right)^{\mathrm{T}} \tag{2.14}$$

and contains the receptive fields of all $|P_T|$ receptor units. The complete filtering transformation from the input image I_S to the vector I_T of receptor responses is

$$I_T = \mathbf{W}^{\mathrm{Rcpt}} \cdot I_S . \tag{2.15}$$

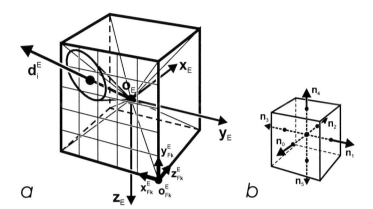

Figure 2.7: Coordinate systems of a cube environment map. *a* Eye (E) and face (F_k) coordinate systems. *b* Normal vectors of the individual cube faces.

2.2.4   Coordinate Systems

The viewing directions of the target receptor units are defined in eye coordinates, whereas the source pixel positions are given in the local coordinate systems of multiple planar images forming the environment map. Thus, a common coordinate system for both source and target pixel locations is required to calculate the receptive fields.

Figure 2.7a shows the eye coordinate system E with origin $\mathbf{o}_E$ and basis vectors $\mathbf{x}_E$, $\mathbf{y}_E$ and $\mathbf{z}_E$. Each face $k \in \{1, \ldots, 6\}$ of the cube environment map has its own coordinate system F_k with origin $\mathbf{o}_{F_k}^E$ and basis vectors $\mathbf{x}_{F_k}^E$, $\mathbf{y}_{F_k}^E$ and $\mathbf{z}_{F_k}^E$, all defined with respect to the eye coordinate system E. $\left\| \mathbf{x}_{F_k}^E \right\|$ and $\left\| \mathbf{y}_{F_k}^E \right\|$ are chosen to equal the width and height of one pixel in the image on the corresponding face k of the environment map. $\mathbf{z}_{F_k}^E = \mathbf{x}_{F_k}^E \times \mathbf{y}_{F_k}^E / \left\| \mathbf{x}_{F_k}^E \times \mathbf{y}_{F_k}^E \right\|$ is set perpendicular to the image plane and normalized to unit length. Thus, the affine transformation from face to eye coordinates of the center vectors $\mathbf{p}^{F_k}$ of the pixels in the image plane is

$$\widehat{\mathbf{p}}^E = \mathbf{M}^{F_k E} \cdot \mathbf{p}^{F_k} + \mathbf{o}_{F_k}^E \qquad (2.16)$$

with the matrix

$$\mathbf{M}^{F_k E} = \left(\mathbf{x}_{F_k}^E, \mathbf{y}_{F_k}^E, \mathbf{z}_{F_k}^E \right) \qquad (2.17)$$

providing scaling and rotation. The normalized vector

$$\mathbf{p}^E = \widehat{\mathbf{p}}^E / \left\| \widehat{\mathbf{p}}^E \right\| \qquad (2.18)$$

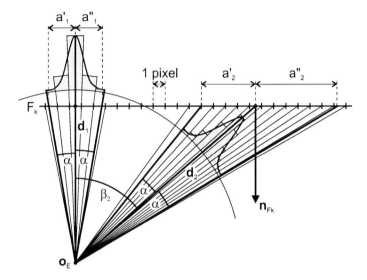

Figure 2.8: Receptive field projection. The sensitivity distribution of the receptive field is projected onto a particular face F_k of the environment map. Although the aperture angle $\Delta\alpha = 2\alpha$ remains constant, the receptive field diameter $\Delta a_i = a_i' + a_i''$ in the texture plane varies for different viewing directions d_i and is minimal when d_i coincides with the face normal n_k. This leads to distorted receptive fields in the texture plane.

represents the direction of a particular pixel in eye coordinates and is used in section 2.2.5 to compute angular distances.

2.2.5   Receptive Field Projection

Each receptor unit q has a specific local viewing direction $d_q^E \in U$ given in eye coordinates. To prevent spatial aliasing due to point sampling, the incoming light intensity is integrated over a conical receptive field around the optical axis d_q^E. Here, each receptive field has an isotropic, Gaussian-shaped sensitivity distribution

$$G_q(\zeta) = \exp(-2\zeta^2/\Delta\rho_q^2) \tag{2.19}$$

with a space-variant half-width angle $\Delta\rho_q$ (Fig. 2.3b). It is defined as a function of the angular distance ζ from the optical axis and has its maximum sensitivity for $\zeta = 0$.

To determine the contribution of each source pixel p to the response of a particular receptive field q the center vectors of all pixels in the environment map are transformed from face to eye coordinates (Eq. 2.16) and normalized to unit length (Eq. 2.18). The angular distance between each resulting pixel direction vector $\mathbf{p}_p^E \in U$ and the optical axis $\mathbf{d}_q^E$ can now be determined using Eq. 2.5, and the corresponding relative sensitivity of the receptive field q in the direction of pixel p is

$$\widehat{w}_{q,p}^{\mathrm{Rcpt}} = \left\{ \begin{array}{ll} G_q\left(\arccos\left(\mathbf{d}_q^E \cdot \mathbf{p}_p^E\right)\right) \cdot A_p, & \mathbf{d}_q^E \cdot \mathbf{p}_p^E > \cos(\frac{1}{2}\Delta\alpha_q) \\ 0, & \text{else} \end{array} \right. . \qquad (2.20)$$

The solid angle A_p covered by the planar, rectangular source pixel p depends on the position on the image plane (Fig. 2.8) and is therefore required as a correction factor. The space-variant aperture angle $\Delta\alpha_q$ (Fig. 2.3b) truncates the Gaussian sensitivity distribution to a conical region around the optical axis. Thus, each receptive field (Eq. 2.13) needs to be normalized using

$$\mathbf{w}_q^{\mathrm{Rcpt}} = \frac{1}{\sum_p \widehat{w}_{q,p}^{\mathrm{Rcpt}}} \widehat{\mathbf{w}}_q^{\mathrm{Rcpt}} \qquad (2.21)$$

to yield a unit gain factor. Fig. 2.9 shows an example for the distorted receptive fields resulting from projecting an isotropic, Gaussian-shaped sensitivity distribution on a cube environment map.

2.2.6   Source Image Size and Resolution

Each receptor unit of the simulated compound eye has a Gaussian-shaped receptive field with maximum sensitivity in the direction $\mathbf{d}$ of its optical axis. The receptive field is limited by an aperture angle $\Delta\alpha$ around the optical axis (Fig. 2.3b), thus it covers a cone-shaped solid angle of the environment. For all directions outside this conical region the sensitivity is set to zero. Since the environment map is composed of planar perspective views, a conical section with one of these images yields an elliptical area of pixels falling within the aperture. These pixels contribute to the output value of the receptor unit and are weighted according to the sensitivity distribution of the receptive field.

When a receptive field is projected onto the face k of the environment map, the number of pixels in the image plane falling within the receptor aperture varies with the eccentricity angle β between the receptor direction $\mathbf{d}$ and the normal vector of the image plane $\mathbf{n}_k$. It is minimal when $\beta = 0$, i.e., when the receptor viewing direction is orthogonal to the image plane (Fig. 2.8). To obtain a minimum diameter of Δa_{min} pixels in the image plane for a given aperture angle $\Delta\alpha$, the faces of the environment map need to have a minimum side length l of

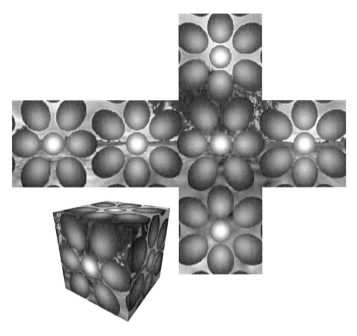

Figure 2.9: Gaussian-shaped receptive fields projected on a cube environment map. For demonstration, large, non-overlapping receptive fields are used in this example.

$$l = \Delta a_{\min} \cdot \frac{\tan(\gamma/2)}{\tan(\Delta\alpha/2)} \qquad (2.22)$$

pixels. The angle γ denotes the aperture of a single face of the environment map, i.e., twice the angle between the face normal n_k and the edge of the image.

2.2.7 Lookup Table Implementation

Each weight vector w_q^{Rcpt} has one entry for each pixel of the input image. This leads to a large but sparsely occupied weight matrix W^{Rcpt}, since each receptive field covers only a small solid angle containing only a small portion of the environment map. An efficient implementation of the weight matrix is achieved by storing only non-zero weights together with the corresponding pixel indices in a pre-computed lookup table

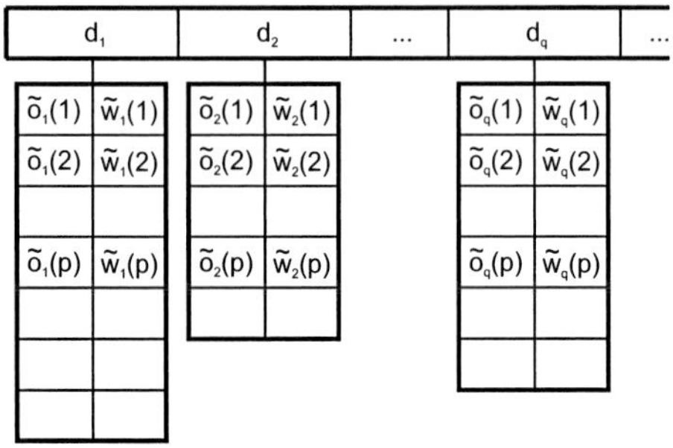

d_1		d_2		...	d_q		...
$\tilde{o}_1(1)$	$\tilde{w}_1(1)$	$\tilde{o}_2(1)$	$\tilde{w}_2(1)$		$\tilde{o}_q(1)$	$\tilde{w}_q(1)$	
$\tilde{o}_1(2)$	$\tilde{w}_1(2)$	$\tilde{o}_2(2)$	$\tilde{w}_2(2)$		$\tilde{o}_q(2)$	$\tilde{w}_q(2)$	
$\tilde{o}_1(p)$	$\tilde{w}_1(p)$	$\tilde{o}_2(p)$	$\tilde{w}_2(p)$		$\tilde{o}_q(p)$	$\tilde{w}_q(p)$	

Figure 2.10: Lookup table (LUT) containing a separate index and weight list for each distorted receptive field.

(LUT). For each receptive field q the set $\tilde{P}_q$ of source pixel indices with non-zero weights is determined by

$$\forall q \in P_T : \tilde{P}_q = \{p \in P_S | w_{q,p}^{\mathrm{Rcpt}} \neq 0\} . \qquad (2.23)$$

With an arbitrary bijective function $\tilde{o}_q : \{1, \ldots, |\tilde{P}_q|\} \to \tilde{P}_q$ ordering the source pixel indices, and a weighting function $\tilde{w}_q : \{1, \ldots, |\tilde{P}_q|\} \to \Re$ defined as

$$\tilde{w}_q(p) := w_{q,\tilde{o}_q(p)}^{\mathrm{Rcpt}} \qquad (2.24)$$

the filtering transformation can be efficiently computed using

$$\forall q \in P_T : I_T[q] = \sum_{p=1}^{|\tilde{P}_q|} \tilde{w}_q(p) \cdot I_S[\tilde{o}_q(p)] \qquad (2.25)$$

to generate the receptor output values I_T from the input image I_S. Fig. 2.10 illustrates the implementation of the index and weighting functions as separate LUTs for each receptor unit q.

2.3   Results

In this section the accuracy and performance of the proposed compound eye simulation model is analyzed. In addition, examples for different spherical receptor distributions are presented, and the virtual environment is shown as seen through each eye configuration.

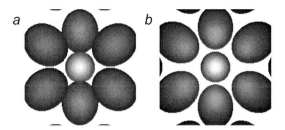

Figure 2.11: Distorted receptive fields on a cube map face. *a*. The elliptical weighted average filter (EWA) shifts the receptive fields towards the image center. *b*. The receptive field projection (RFP) proposed here yields the correct distortion.

2.3.1 Accuracy

In the compound eye simulation model presented here the currently visible environment is represented on multiple planar perspective views rather than on a single spherical image. Thus, to determine the light intensity or color seen by a particular receptor unit, the receptive field needs to be adjusted to compensate for perspective distortions in the image plane. The elliptical weighted average filter (EWA) suggested by Greene and Heckbert (1986) provides an approximate solution since it distorts the originally circular filter masks into elliptical areas on the image plane, using a concentric weight distribution around the center of each ellipse. Fig. 2.11*a* shows the resulting distorted sensitivity distributions of isotropic, Gaussian-shaped receptive fields on a cube environment map. In contrast, the filtering method proposed here projects the receptive fields onto the planar faces of the environment map, resulting in anisotropic distorted weight distributions as depicted in Fig. 2.11*b*.

For both approaches the maximum sensitivities of the distorted receptive fields are located at the same positions in the image plane. However, the EWA filter (Fig. 2.11*a*) shifts the surrounding weight distribution towards the image center, thereby changing the receptor viewing directions. For large aperture and acceptance angles this leads to considerable deformations of the eye geometry. These errors do not occur in the receptive field projection method presented here (Fig. 2.11*b*).

The accuracy of both filtering algorithms can be assessed by comparing the centroid directions of the distorted receptive fields with the original optical axes of the receptor units. The normalized centroid direction vector b_q^E of receptor unit q in eye coordinates is

$$b_q^E = \frac{\sum_{p \in \widetilde{P}_q} \widetilde{w}_q(p) \cdot \mathbf{p}_q^E(p)}{\| \sum_{p \in \widetilde{P}_q} \widetilde{w}_q(p) \cdot \mathbf{p}_q^E(p) \|} . \tag{2.26}$$

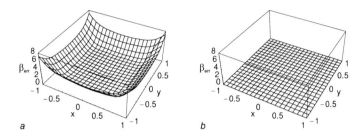

Figure 2.12: Directional error of distorted receptive fields as a function of the intersection point of the receptor viewing direction with a cube map face. The error angle β_{err} [°] describes the deviation of the centroid direction vector of a distorted receptive field from the original viewing direction of the corresponding receptor unit. The isotropic, Gaussian-shaped receptive fields have an aperture angle of $\Delta\alpha = 30°$, an acceptance angle of $\Delta\rho = 20°$, and a minimum diameter of $\Delta a_{\mathrm{min}} = 20$ pixels. *a.* For the elliptical weighted average filter (EWA) the error increases with the eccentricity of the viewing direction from the center of the image plane. *b.* The receptive field projection (RFP) preserves the intended viewing direction over the entire cube environment map.

$\mathbf{p}_q^E(p)$ is the normalized center vector of the environment map texture pixel p in eye coordinates. It is obtained using Eq. 2.16 and Eq. 2.18. $\tilde{w}_q(p)$ is the weight assigned to pixel p of receptive field q (Eq. 2.24) with $\sum_{p \in \tilde{P}_q} \tilde{w}_q(p) = 1$.

A measure for the deviation of the centroid direction vector $\mathbf{b}^E$ from the original viewing direction $\mathbf{d}^E$ is the error angle

$$\beta_{\mathrm{err}} = \arccos(\mathbf{b}^E \cdot \mathbf{d}^E). \tag{2.27}$$

Fig. 2.12 compares both filtering algorithms and shows the directional error β_{err} of an isotropic, Gaussian-shaped receptive field as a function of the intersection point of the optical axis with a face of a cube environment map. The receptive field has an aperture angle of $\Delta\alpha = 30°$, an acceptance angle of $\Delta\rho = 20°$, and a minimum diameter of $\Delta a_{\mathrm{min}} = 20$ pixels on the image plane. For the elliptical weighted average filter (EWA) the error increases with the eccentricity of the optical axis from the center of the image plane (Fig. 2.12*a*). This leads to considerable distortions in the arrangement of optical axes, in particular when receptive fields with large aperture angles are used. For a receptor unit with an optical axis oriented towards a corner of the cube environment map, the centroid direction of the resulting distorted sensitivity distribution deviates from the original viewing direction by $\beta_{\mathrm{err}} = 7.2°$.

In contrast, the receptive field projection method proposed here ensures that the centroid direction vectors of all receptive fields remain coincident with the correspond-

$|P_T|$	$\Delta\varphi$ [°]	$\Delta\rho$ [°]	$\Delta\alpha$ [°]	Δa_{min} [pix]	CubeEM Size [pix]		RF Size [pix]		
					face width	total #pix	min	mean	max
642	8.6	10.0	20.0	3	20	2400	10	18.0	32
				6	36	7776	32	58.9	107
2562	4.3	5.0	10.0	3	36	7776	4	14.6	34
				6	72	31104	30	58.9	126
10242	2.2	2.5	5.0	3	72	31104	4	14.7	35
				6	140	117600	28	55.9	136

Table 2.1: Size of the cube environment map and the receptive fields for different eye configurations. $|P_T|$ is the number of receptor units, $\Delta\varphi$ is the inter-receptor angle. Each receptive field has an acceptance angle $\Delta\rho$, an aperture angle $\Delta\alpha$, and a minimum diameter of Δa_{min} pixels on the environment map.

ing receptor axes. Thus, angular errors are completely avoided (Fig. 2.12*b*) and all optical axes are preserved independently of their particular viewing direction or receptive field size.

2.3.2 Performance

The compound eye simulation model was implemented and tested on a standard PC with a 450 MHz Intel Pentium II CPU and an Nvidia GeForce2 GTS graphics processor. Input images were generated using an OpenGL-based computer simulation of a 3D virtual world.

Three eye configurations with 642, 2562 and 10242 receptor units were tested using two different minimum diameters Δa_{min} in the image plane. For each of the six resulting cube environment maps the width of a single face as well as the total number of pixels are shown in Tab. 2.1. Further, the minimum, mean and maximum numbers of pixels contained in the receptive fields are presented for each eye configuration. The data demonstrate that the anti-aliasing quality indicated by the receptive field size can be maintained in the same range for different spatial resolutions of the simulated eye by using a constant minimum diameter Δa_{min}. However, higher resolutions defined by a larger number of receptor units $|P_T|$ and a smaller inter-receptor angle $\Delta\varphi$ require larger environment maps.

Tab. 2.2 shows the performance of the LUT-based compound eye simulation for a gray value and an RGB color version of the cube environment map, using the same receptive field parameters as in Tab. 2.1. As expected, the filtering algorithm achieves the highest update rates for gray value images and small receptor numbers, and is slowest for high resolution eyes and color images. In addition to the isolated filtering

$|P_T|$	$\Delta\varphi$	$\Delta\rho$	$\Delta\alpha$	Δa_{min}	Gray [s^{-1}]		Color [s^{-1}]	
	[°]	[°]	[°]	[pix]	F	RCF	F	RCF
642	8.6	10.0	20.0	3	2208.1	11.0	1143.5	11.6
				6	711.2	10.7	367.8	11.2
2562	4.3	5.0	10.0	3	374.3	10.4	246.1	11.3
				6	107.9	7.7	67.8	9.7
10242	2.2	2.5	5.0	3	83.8	7.7	55.9	9.7
				6	27.3	3.9	16.1	6.3

Table 2.2: Performance of the compound eye simulation. The LUT-based filtering algorithm proposed here is tested on a standard PC with Intel Pentium II 450 MHz CPU and Nvidia GeForce2 GTS graphics processor, using the eye configurations described in Tab. 2.1. The update rates are given for both gray value and RGB color images, and are shown separately for the isolated filtering operation (F) and for the complete simulation including rendering the example scene, copying the image to CPU memory and filtering (RCF).

operation the performance of the full simulation is shown, including rendering the example scene to a cube environment map (Fig. 2.6), copying the resulting images from the frame buffer to CPU memory, and filtering. Evidently, the filtering algorithm is very efficient and requires only a small fraction of the total computation time. In the full simulation the update rates are higher for color images than for gray values since the latter cannot be generated directly by the OpenGL-based simulation and need to be converted from RGB color mode in an additional processing step.

2.3.3   Example Eye Configurations

Provided monocular vision, i.e. a singular center of projection for all optical axes, the compound eye simulation model permits arbitrary spherical receptor arrangements. Three examples for different spherical eye configurations are presented in Fig. 2.13. Fig. 2.13a shows a minimalistic arrangement of $|P_T| = 225$ receptor units with an angular distance of $\Delta\varphi = 6°$, covering only selected regions of the sphere. This arrangement was used in preliminary studies to demonstrate basic visual orientation strategies for flight control (Neumann and Bülthoff, 2000, 2001). A longitude-latitude graticule composed of $|P_T| = 800$ receptor units with an angular spacing of $\Delta\varphi = 9°$ along the directions of azimuth and elevation is depicted in Fig. 2.13b. Here, the inter-receptor angle $\Delta\varphi$ varies between $9°$ at the horizon and $0°$ at the poles. Fig. 2.13c shows a hexagonal grid with $|P_T| = 642$ receptor units and an approximately position-invariant angular distance of $\Delta\varphi = 8.6°$ between all adjacent local viewing directions. Spherical hexagonal grids are comprehensively described in Chapter 3 and are used

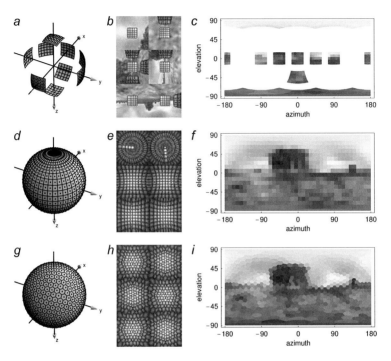

Figure 2.13: Example receptor configurations. *a,b,c* Minimalistic eye model covering only selected regions of the sphere ($|P_T| = 225$, $\Delta\varphi = 6°$). *d,e,f* Longitude-latitude graticule with $9°$ spacing ($|P_T| = 800$, $\Delta\varphi \in [0°, \ldots, 9°]$). *g,h,i* Hexagonal grid ($|P_T| = 642$, $\Delta\varphi = 8.6°$). The left column shows the omnidirectional distribution of receptor directions. The six faces of the cube environment map containing the distorted, overlapping Gaussian-shaped receptive fields are depicted in the center column. The right column shows the example scene (Fig. 2.6) as seen through each eye model.

for image acquisition in Chapter 4.

In all eye configurations anti-aliasing is ensured by overlapping, Gaussian-shaped receptive fields. Their distorted projections on the six faces of a cube environment map are depicted in the center column of Fig. 2.13. In the longitude-latitude grid and in the hexagonal arrangement the receptive fields cover the entire spherical field of view. The right column of Fig. 2.13 shows Mercator projections of the example scene (Fig. 2.6) as seen through each eye model.

2.3.4 Insect-Inspired Receptor Distributions

Insect-inspired spherical receptor distributions and the resulting omnidirectional images are shown in Fig. 2.14. The relative local receptor densities are modeled after biological examples (Fig. 2.4) given by Land (1997b). As in the previous section, the resulting spherical images show the example scene (Fig. 2.6) as seen through each eye model. They are depicted as distorted Mercator projections with a full $360° \times 180°$ field of view. An eye configuration with an increased resolution in the equatorial region is depicted in Fig. 2.14b. It magnifies details around the horizon which are not visible in the homogeneous receptor distribution shown for comparison in Fig. 2.14a. Similar eyes are used by animals living in flat environments such as desert ants, water striders or empid flies. Fig. 2.14c shows a receptor distribution with a frontal acute zone as in male house flies or blow flies, emphasizing the center region of the image. The configuration depicted in Fig. 2.14d resembles the eyes of dragonflies. It exhibits an overall increased resolution as well as double acute zones in the frontal and the frontodorsal regions. These regions appear strongly magnified in the resulting image.

2.4 Discussion

In this chapter I presented an algorithm that generates low resolution omnidirectional images from multiple perspective views of the environment. All perspective views are taken from a single viewpoint and are arranged as a cube environment map covering the entire visible surroundings. A Gaussian-shaped sensitivity distribution around each local sampling direction suppresses high spatial frequencies in the omnidirectional image and inhibits spatial aliasing. The Gaussian filter masks are projected onto the environment map in order to compensate for perspective distortions. The compound eye simulation model is both accurate and efficient. It allows to specify arbitrary, space-variant distributions of local sampling directions in a spherical field of view, and is capable of reconstructing an insect's view in highly realistic virtual environments at high frame rates.

2.4.1 Environment Mapping

Compared to previous approaches using ray tracing to simulate insect vision, the environment mapping method presented here offers a number of advantages. Most importantly, it is very efficient since it makes use of current 3D computer graphics hardware which is highly optimized for generating planar perspective images of complex scenes at high frame rates. High update rates are crucial for the optimization of large populations by online learning or genetic algorithms (Huber et al., 1996; Neumann et al.,

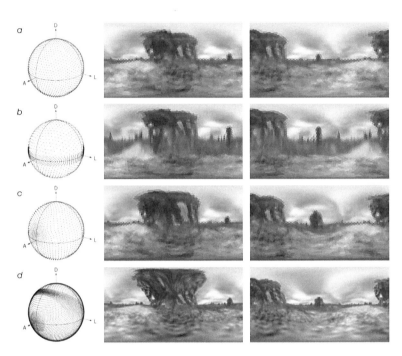

Figure 2.14: Space-variant spherical vision with different insect-inspired eye models (cf. Fig. 2.4). The left column depicts the different spherical receptor distributions (A:anterior, L:lateral, D:dorsal), the right column shows the example scene (Fig. 2.6) as seen through each eye model as distorted Mercator projections with a full $360° \times 180°$ field of view. *a*. Homogeneous receptor distribution. *b*. Increased resolution along the horizon as in desert ants, water striders or empid flies. *c*. Frontal acute zone as in male house flies or blow flies. *d*. Overall increased resolution and double acute zones in the frontal and the frontodorsal regions as in dragonflies.

1997; Dale and Collett, 2001), and realistic input stimuli are inevitable to evaluate visual control models for both animals and real-world robots. Recent studies emphasize the role of natural image statistics for visual information processing in insects (Dror, O'Carroll, and Laughlin, 2001). In contrast, many ray tracing-based simulations use extremely simplified visual stimuli (Cliff, 1991; Mura and Franceschini, 1994; Giger, 1995) since realistic scenes would require enormous computational resources or cannot be computed in reasonable time. As a further way to reduce computation time, most ray tracing-based studies use only one up to five discrete sampling directions for each receptor unit. With realistic input stimuli this would lead to aliasing effects that strongly impair subsequent processing stages such as local motion detection. In contrast, environment mapping provides highly accurate spatial low pass filtering of the visual input since each local receptive field integrates over a large number of pixels. Perspective images of sufficient size can be efficiently computed.

Cube environment mapping was originally developed for computer graphics as an efficient method to approximate reflections on shiny objects (Greene, 1986) and to generate high resolution Omnimax frames from perspective images (Greene and Heckbert, 1986). For both applications the elliptical weighted average filter (EWA) was suggested to compensate for perspective distortions although it gives rise to systematic errors in the local sampling directions (Section 2.3.1). In many situations, these errors can be neglected: Human observers experience a reflective object even for large deviations from the physically correct reflection, and the small filter masks required for high resolution Omnimax frames permit only small angular errors of the local viewing directions. However, the EWA filter would lead to severe distortions in the arrangement of local receptor axes when used with the large acceptance angles required for low resolution compound eye simulations. Therefore, the approximate EWA filter is best replaced by the accurate receptive field projection (RFP) method proposed here.

Defining the filter masks directly on the sphere eliminates many problems that occur when spherical images are filtered in planar representations. For instance, Mercator projections are widely used to store omnidirectional images in a single, planar rectangular image (Franz, Neumann, Plagge, Mallot, and Zell, 1999; Tammero and Dickinson, 2002). However, pixels of equal shape and size in the Mercator plane are unequally distorted when mapped back onto the sphere, as demonstrated in Fig. 2.13*d* and *f*. The distortion increases with the angular distance from the equator and leads to singularities at the poles. Thus, standard space-invariant filtering in the image plane is space-variant on the sphere and can only be used as an approximate solution as long as the visual stimulus is restricted to regions around the equator. In contrast, receptive field projection provides full control of both shape and sensitivity distribution of each receptive field independently of the specific viewing direction or local receptor density.

It is worth noting that in principle the perspective images on the faces of the environment map need to be updated only for translatory self-motion of the eye point.

For pure rotations, the environment map may remain unchanged and the receptive fields could be rotated instead. However, this would require a computationally costly re-calculation of the entire lookup table for each rotation step since the lookup table has a fixed orientation with respect to the environment map. Therefore, the filtering algorithm presented here is most efficient when the LUT is determined once and the environment map is updated for each simulation step. This has the additional advantage that all object movements or other changes in the environment are immediately visible to the simulated compound eye.

2.4.2 Real-World Implementation

The spherical filtering algorithm proposed here can be used with any source of planar perspective images, including polygon-based computer graphics and ray tracing, as well as real-world cameras. In a real world implementation, the perspective images on the faces of an environment map can be acquired by a cluster of cameras, one for each face. A tetrahedral arrangement of cameras may be preferred over a cube configuration although tetrahedron environment maps contain more pixels and require larger filter masks than cube maps of comparable resolution (Appendix A). However, the tetrahedral arrangement requires only four instead of six cameras, which constitutes the minimum number of planes required to enclose a point in 3D space.

As pointed out by Swaminathan and Nayar (2000), camera clusters are likely to have nonsingular viewpoints, and the wide angle lenses required in a cube or tetrahedron camera arrangement cause further distortions of the image. To obtain an environment map composed of planar perspective images these distortions need to be removed before the omnidirectional sampling and filtering is applied. This requires an appropriate image warping operation which is usually implemented as an additional lookup table. However, this intermediate processing step is not needed if the receptive fields are mapped directly onto the raw camera images using the lens distortion parameters known from camera calibration. This yields a single, combined lookup table for both the inverse camera distortion and the spherical sampling of the input images. Thus the same efficient algorithm can be applied to computer generated perspective images in virtual environments and to wide angle camera clusters in the real world.

2.4.3 Applications

Applications of this approach can be envisioned for both modeling the visual information processing in insects and for the development of novel, biomimetic vision systems. Both biological and biomimetic studies benefit from recent developments in computer graphics and virtual reality technology, providing photorealistic visual stimuli and realistic models of body and world physics (Terzopoulos and Rabie, 1995). Thus, simulations are no longer restricted to simplified or idealized input data, but provide test environments comparable to real-world robotics. In contrast to robots, simulations are easier to maintain, are not limited to available sensor and effector hardware, and allow full access and control of all aspects of the action-perception cycle including all parameters of body, world, and information processing.

In biological studies the eye simulation model can be used to accurately reconstruct the visual stimulus as it is perceived by an insect during an experiment in tethered or free flight. This requires that the distribution of the receptor viewing directions in the original insect eye is known, that a realistic simulation model of the visual stimulus or the environment is provided, and that the physiological or behavioral responses such as the depolarization of specific interneurons or the flight trajectory of the animal are recorded with sufficient accuracy. This approach allows to derive models of visual information processing and behavior control in insects from correlations of the reconstructed input stimulus with the recorded responses of the animal. In addition, processing and control models can be implemented as part of the simulation, allowing them to be tested and evaluated in the same open- and closed-loop experiments as the animals. The observed physiological and behavioral responses can then be directly compared.

A further application of the proposed compound eye simulation model is the development of novel, insect-inspired computer vision systems for tasks like visual self-motion control and navigation. These systems are expected to be simpler, more robust and more efficient than existing technical solutions which are easily outperformed by flying insects in spite of the low visual acuity and small brain size of the animals. For instance, the eye model proposed here is used in Chapter 4 to investigate visual orientation strategies and optimal receptive fields for biomimetic flight control.

Chapter 3

Insect-Inspired Spherical Image Processing

In this chapter I propose geodesic grids as innovative data structures for the representation and processing of discrete spherical images. The gray or color values of individual pixels reside on a hexagonal lattice formed by the vertices of a recursively subdivided icosahedron, resulting in a quasi-homogeneous distribution of sampling directions on the viewing sphere. This avoids polar singularities and space-variant distortions of pixel shape and size which are inherent to conventional, planar representations of spherical images in Cartesian or polar coordinates. Directional information such as local optic flow is represented on the edges of the grid. The underlying graph structure of geodesic grids permits an efficient implementation of various local image processing operations such as spatial low pass and high pass filtering by spherical convolution, the computation of spherical resolution pyramids, temporal filtering, and local motion detection. This type of local processing on low resolution spherical images with a hexagonal pixel arrangement closely resembles the computations performed by the retinotopic layers in the visual system of insects. Therefore, geodesic grids provide an excellent basis for both modeling insect vision in biological studies and for developing innovative, biomimetic approaches in omnidirectional machine vision.

3.1   Introduction

Omnidirectional vision plays a central role among the various sensory modalities controlling the spatial behavior of flying insects. Despite the low resolution of their compound eyes, ranging from 700 receptor units per eye in the fruit fly *Drosophila melanogaster* up to 28000 in the dragonfly *Anax junius* (Land, 1997b), flying insects exhibit remarkably fast and robust visual control strategies for a multitude of highly sophisticated three-dimensional flight maneuvers such as high speed pursuit of prey or mating partners (Srinivasan and Bernard, 1977; Collett, 1980a; Kirschfeld, 1997), avoiding obstacles during flight (Tammero and Dickinson, 2002; Kelber and Zeil, 1997), and hovering on the spot (Collett and Land, 1975a; Collett, 1980b; Zeil and Wittmann, 1989; Kelber and Zeil, 1990). With a brain smaller than a cubic millimeter this requires an extremely efficient and highly specialized processing of the omnidirectional views perceived during flight. A major part of these computations takes place in the early layers of the insect visual system which are crucial for the local selection and pre-processing of behaviorally relevant features of the input stimulus. In particular, they perform local operations such as noise reduction, contrast enhancement and local motion detection in a massively parallel fashion simultaneously on the entire spherical input image.

In this chapter I introduce novel, insect-inspired data structures and algorithms for the efficient and homogeneous representation and processing of discrete spherical images. In biological studies, they can be used to implement computational models of the retinotopic layers in the insect visual system, allowing to reconstruct the neural images resulting from the first processing stages directly on the sphere. In biomimetic computer vision systems, the insect-inspired approach provides a highly efficient alternative to conventional methods for spherical image processing. It is therefore particularly useful for applications such as visual flight control and navigation.

3.1.1   Omnidirectional Visual Processing in Flies

The visual system of the fly consists of the ocelli, the compound eyes, and the optic lobes (Strausfeld, 1976). Each compound eye is composed of discrete receptor units sampling about one hemisphere of the environment. The photoreceptors convey their output signals from the retinae into the optic lobes which are composed of four successive visual neuropils, the lamina, the medulla, the lobula and the lobula plate. Anatomical studies show that these neuropils occupy a large part of the fly brain and are organized in a columnar, retinotopic fashion (Strausfeld, 1989), i.e., the signals from neighboring receptor units are represented at neighboring locations throughout subsequent layers. This indicates massively parallel processing of the incoming spherical images, involving mainly local interactions between directly adjacent units of the columnar array (Hausen and Egelhaaf, 1989). In the following I briefly review the different stages of spherical image processing in the fly visual system.

Compound Eyes

Insect compound eyes sample the surrounding environment in an approximately spherical field of view, using a low resolution hexagonal array of discrete receptor units (cf. Section 2.1.1). Neighboring ommatidia receive their visual input from partially over-

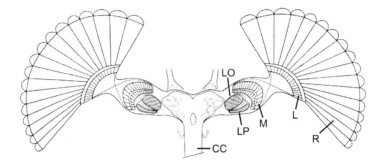

Figure 3.1: Visual system of the blowfly *Calliphora erythrocephala*. Each retina (R) is connected to an optic lobe composed of the successive, retinotopic processing layers lamina (L), medulla (M), lobula (LO) and lobula plate (LP). The lobula complex projects through the cervical connective (CC) to the thoracic-abdominal ganglia (not shown) responsible for motor activation. (Modified, with permission, from N.J. Strausfeld: Beneath the compound eye: Neuroanatomical analysis and physiological correlates in the study of insect vision. In D.G. Stavenga and R.C. Hardie (Eds.): *Facets of Vision*, p. 321, Fig. 1. © 1989 Springer-Verlag.)

lapping, Gaussian-shaped receptive fields around their optical axes (Götz, 1965; Land, 1997b). This eliminates spatial aliasing due to high spatial frequencies which cannot be resolved by the sampling grid. Thus, spatial low pass filtering by compound eye optics constitutes the first stage of spherical image processing in the insect visual system. Note that the sampling grid, i.e., the spatial arrangement of receptor units in the compound eye, defines the topological layout of all subsequent, retinotopic processing layers. Since each eye samples about one hemisphere of the environment with small or no overlap, subsequent image processing in the optic lobes is also performed on hemispherical images.

Retina

Each ommatidium in the fly compound eye contains a well-organized set of eight photoreceptors. Two receptors are located in the center of the ommatidium and share the same optical axis, whereas each of the six surrounding receptors shares its viewing direction with the central receptors in one of the six directly adjacent ommatidia. The axons of all photoreceptors oriented towards the same point in space converge in the lamina. This arrangement is known as neural superposition (Kirschfeld, 1967) and enhances the photon efficiency of the system at least six fold (Nilsson, 1989). Using multiple receptors for the same viewing direction yields a significant improvement of the signal-to-noise ratio at low light levels without sacrificing spatial acuity. In addition to visual input acquisition, the retinal photoreceptors have at least two important functions in the local processing of the incoming spherical images: temporal low pass filtering, and adaptation to the ambient light level.

Temporal low pass filtering is required since the neural processing speed in the fly visual system is limited by a time constant of 0.5 ms for the hyperpolarization component of synaptic transmission (Laughlin, Howard, and Blakeslee, 1987). In analogy to spatial aliasing, faster changes of the input signal may lead to artifacts and increase the noise level. To ensure that synaptic transmission is faster than the incoming signal changes, high temporal frequencies are suppressed directly in the receptor using a time constant of at least 2 ms for phototransduction (Howard, Blakeslee, and Laughlin, 1987). A further, local image processing operation performed by the retina is the compression of the wide range of diurnal light intensities over five orders of magnitude (Laughlin, 1989) to the limited response range of the photoreceptors. As in many other animals, the photoreceptors of the fly are capable of adapting their sensitivity to the ambient light level, resulting in an approximately logarithmic transformation on the optical signals (Zettler, 1969).

Lamina

The lamina is a retinotopic array of repeated subunits, the lamina cartridges. There is one cartridge for each ommatidium, receiving axons from all retinal photoreceptors sharing the same viewing direction. Each cartridge contains two large monopolar cells (LMCs) which constitute the main processing units in the lamina and project retinotopically to the medulla.

The response characteristics of lamina LMCs are well known from electrophysiological studies. LMCs remove the local average background intensity from the signal by coding the spatiotemporal contrast, i.e., the spatial and temporal changes of the signal (Srinivasan, Laughlin, and Dubs, 1982; Laughlin, 1989). The spatial receptive field of an LMC is therefore organized in a center-surround fashion, subtracting the local average over a number of neighboring units from the signal of the center unit. The temporal receptive field exhibits a similar layout. Here, the temporal average over a certain, immediately preceding period of time is subtracted from the current signal. This operation is also known as predictive coding (Srinivasan et al., 1982) since a constant or slowly changing background intensity can be predicted from the context and therefore contains no additional information. Thus, the background intensity is redundant and can be removed from the signal. In addition to removing very low spatial and temporal frequencies, very high frequencies also need to be suppressed since they are corrupted by receptor and transduction noise. A theoretical model predicting the spatiotemporal receptive fields of LMCs is presented by van Hateren (1992).

Finally, LMCs amplify their output signals in order to make full use of the limited response range. This operation is necessary since new noise is added during the transduction to subsequent layers. The amplification maximizes the signal-to-noise ratio, allowing to distinguish more levels in the signal (Laughlin, 1989).

Medulla

The medulla continues the regular, retinotopic organization of the preceding processing layers (Strausfeld, 1976). It is assumed to be involved in local motion detection, and there is evidence for a variety of small field nondirectional and directional motion sensitive units in the medulla (Strausfeld, 1989).

A scheme for local motion detection in the insect visual system, the so-called Reichardt detector, was derived from behavioral experiments with insects (Hassenstein and Reichardt, 1956; Reichardt, 1969; Egelhaaf and Borst, 1993). Although the exact neuronal implementation and structure of this elementary motion detector (EMD) has not yet been identified (Borst and Haag, 2002; Hausen and Egelhaaf, 1989), its computational properties are well understood (Poggio and Reichardt, 1973; Borst and Egelhaaf, 1989, 1993; Mallot, 2000). A Reichardt-type EMD receives two intensity signals from separate locations in the input image and detects a one-dimensional component of local image motion between these two input points. In each of two mirror-symmetrical semidetectors an asymmetrical phase shift is introduced to the input signals by different temporal low pass filtering on both sides, thereby delaying one of the signals more than the other. The subsequent correlation of both signals is maximal when the image displacement during the temporal delay equals the angular distance between the two sampling points, and smaller for both slower and faster image motion. No response is obtained for image motion in the opposite direction. Subtracting both semidetector responses yields the full detector output signal indicating the presence and direction of local image motion.

The output signal of a single EMD depends on the phase, contrast, and spatial frequency content of the input pattern. In particular, low spatial frequencies cause transients and oscillations in the output signal. However, the dependence on the input pattern can be reduced by spatial integration over an array of multiple EMDs looking at different phases of the stimulus pattern (Borst and Haag, 2002). This wide-field integration takes place in the lobula plate.

Lobula Complex

The lobula complex is composed of two retinotopically organized structures, the anterior lobula and the posterior lobula plate (Strausfeld, 1976). Both contain numerous different cell types which may contribute to local motion detection. The lobula plate receives input from columnar cells in the medulla and the lobula.

In contrast to the strictly local image processing in the preceding layers, the lobula plate contains about 60 so-called tangential neurons which respond in a directionally selective manner to wide-field motion stimuli (Hausen, 1984; Hausen and Egelhaaf, 1989). Each tangential neuron can be individually identified by its response characteristics, as well as by the particular shape of its dendritic tree which integrates local motion signals over large parts of the visual field. Lobula plate tangential cells can be grouped into different subclasses according to their preferred orientation. Cells of the horizontal system (HS-cells) respond preferentially to wide-field horizontal image motion (Hausen, 1982a, 1982b), whereas units of the vertical system (VS-cells) are mainly sensitive to vertical motion (Hengstenberg, Hausen, and Hengstenberg, 1982; Hengstenberg, 1982). Detailed investigations showed that the receptive fields of most tangential cells are organized in a highly complex fashion closely resembling specific, self-motion-induced optic flow fields. For instance, each VS neuron is sensitive to rotatory optic flow about a particular axis rather than to simple vertical motion (Krapp and Hengstenberg, 1996; Krapp et al., 1998). This lead to the assumption that tangential cells act as neuronal matched filters for specific patterns of image motion (Franz and Krapp, 2000; Krapp, 2000).

In essence, the insect visual system processes discrete spherical images in various successive, retinotopic layers. Operations such as spatiotemporal band pass filtering and local motion detection are applied to each location in the image. In the next section I show why traditional methods for spherical image processing in computer vision are not suitable to implement an efficient simulation model of the insect visual system.

3.1.2   Spherical Image Processing in Machine Vision

The growing interest in omnidirectional imaging systems for surveillance, immersive multimedia, as well as robot vision and navigation has lead to the development of various methods to capture panoramic or spherical images of the environment. Examples include multi-camera arrangements (Swaminathan and Nayar, 2000), rotating cameras (Nayar and Karmarkar, 2000), wide-angle lenses, and catadioptric systems combining lens optics with reflective surfaces (Chahl and Srinivasan, 1997; Nayar and Boult, 1998). Since most of these systems are based on conventional video cameras, the resulting omnidirectional images are captured by planar CCD or CMOS sensors, and stored in computer memory as rectangular arrays of rectangular or square pixels.

Planar Representations of Spherical Images

The most common planar representations of spherical images, as well as their projections on the sphere are illustrated in Fig. 3.2. The Mercator plane (Fig. 3.2a) provides a Cartesian coordinate system of azimuth and elevation. However, pixels of apparently

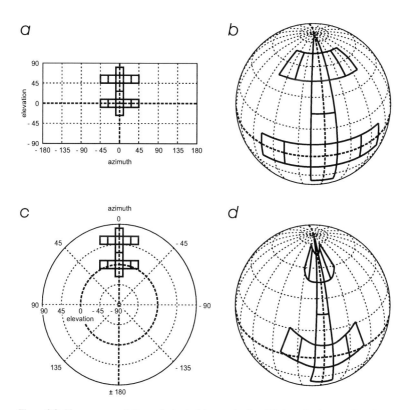

Figure 3.2: Planar representations of spherical images (*a,c*) and their projections on the sphere (*b,d*). Exemplary pixels are indicated by thick black lines. Square pixels of apparently uniform shape, size and distance in the Mercator plane (*a*) are severely distorted on the viewing sphere (*b*). The resolution is lowest at the equator and increases towards the poles where the image becomes singular. The polar coordinates of the catadioptric imaging plane (*c*) introduce even more complex pixel distortions (*d*). Here, the region of lowest resolution is located at the pole mapped to the center of the planar image, and increases towards the opposite pole, again introducing a singularity. In addition, pixels change their local orientation with respect to the polar coordinate system. Both representations are not suitable for the homogeneous representation and processing of spherical images.

uniform shape, size and distance in the Mercator plane are severely distorted on the sphere (Fig. 3.2b). In particular, the resolution, i.e., the number of pixels per unit solid angle, is lowest at the equator and increases towards the poles. At the poles the image becomes singular, and all pixels with elevation angles of $\pm 90°$ are mapped to a single point on the sphere. The pixel boundaries as well as the lines connecting the centers of adjacent pixels are aligned with the Cartesian coordinate system. Lines with constant azimuth, e.g., the vertical boundaries of pixels, are located on great circles of the sphere. Note that the distortions are less pronounced for small absolute elevation angles, and may be tolerable for an approximately homogeneous representation and processing of panoramic images covering only a narrow band around the equator.

Polar coordinates (Fig. 3.2c) provide an alternative method to represent spherical images in the plane by using azimuth and elevation as polar angle and radius, respectively. However, the polar representation causes even more severe distortions of the original spherical image than the Mercator projection. Here, the image resolution is lowest in the region around the pole located in the center of the planar image, and increases towards the opposite pole which is represented by the entire outer boundary of the disc-shaped planar image. All pixels located on the boundary circle are mapped to a single point on the sphere, thereby causing a polar singularity. As above, apparently uniform pixels in the planar image exhibit space-variant distortions of shape, size, and inter-pixel distance on the sphere. In addition, the rectangular grid of pixels in the image plane is not aligned with the polar coordinate system, causing space-variant changes of pixel orientation. This gives rise to further distortions, as illustrated in (Fig. 3.2d).

Polar representations are obtained from catadioptric imaging systems composed of a standard lens camera pointing towards the apex of a conical or parabolic mirror. Note that these systems generate panoramic images with large elevation angles, but usually the field of view does not cover the entire sphere due to blind zones at the poles. For instance, in a system using a reflective surface mounted above an upwards pointing camera, the south pole of the viewing sphere (elevation = $-90°$) is not visible due to self-imaging of the camera, whereas the north pole (elevation = $90°$) is occluded by the mirror. Further, a particular shape of the curved reflector is required to achieve a linear distribution of elevation angles along the radius of the circular image (Chahl and Srinivasan, 1997). Other shapes such as reflective cones (Franz, Schölkopf, Mallot, and Bülthoff, 1998b) or spheres (Röfer, 1997) introduce additional, radial distortions of the image along the elevation axis.

Spherical Image Processing in the Plane

Standard operations in digital image processing include, e.g., image smoothing or contrast enhancement by convolution with appropriate low pass or high pass filter kernels, as well as various algorithms for the computation of optic flow from image sequences. These and innumerable other operations are extensively described in textbooks on digital image processing (Jähne, 2001). Most of the operations are defined in a two-dimensional, Cartesian coordinate system, i.e., they are designed and optimized for the processing of planar images stored in rectangular arrays of rectangular pixels. However, applying them directly to planar representations of spherical images such as Mercator or catadioptric projections results in homogeneous filtering only in the imaging plane, and leads to severe, space-variant distortions on the sphere. For instance, a two-dimensional convolution kernel of apparently constant size and shape in the Mercator or catadioptric imaging plane becomes highly distorted on the sphere (Fig. 3.2), resulting in space-variant filtering of the actual spherical image.

Some recent studies compensate for the distortions by using space-variant processing in the imaging plane to achieve homogeneous filtering on the sphere. Daniilidis, Makadia, and Bülow (2002) define filter kernels in the spherical domain but perform the actual filtering operation in the catadioptric plane where the image information resides, thus the filter masks need to be distorted in a space-variant manner. Chahl and Srinivasan (2000) use catadioptric image acquisition and representation for panoramic images with an almost spherical field of view. Prior to processing, they unwarp the image onto the Mercator plane where the filtering takes place. However, the image is still distorted in the Mercator plane, again requiring space-variant filtering to compensate.

In conclusion, planar representations are not suitable for the homogeneous processing of spherical images since they require space-variant filtering to match the distortion of the image. In addition, the resolution of homogeneous filtering is restricted to the lowest resolution occurring in the planar image, i.e., at the equator of the Mercator plane and at the central pole of the catadioptric plane. Thus, the higher resolution in all other image regions containing more pixels per unit solid angle is wasted and slows down processing since it requires larger filter masks. Even purely local operations which are applied to individual pixels, such as signal amplification or temporal filtering, are inefficient due to the over-representation of most image regions. Further, the circular image required for the polar representation needs to be stored in computer memory as a rectangular or square array of pixels, resulting in large, unused regions in the corners of the image. And finally, a true spherical image is seamless and cyclic in all directions, whereas planar representations have borders. Filter masks exceeding a border need to be divided and applied to multiple non-contiguous parts of the planar image.

3.1.3 Modeling the Insect Visual System

Many studies on biologically inspired spatial behavior in artificial systems (cf. Franz and Mallot (2000) for a review on biomimetic robot navigation) are based on simplified models of the insect visual system, and include omnidirectional image acquisition as well as visual motion detection by insect-inspired, correlation-type EMDs. Since most of these studies use ground-based robots with only two or three degrees of freedom, i.e., rotation about the vertical axis and translation in the two-dimensional plane, omnidirectional image acquisition and processing can be restricted to a panoramic, ring-shaped view of the horizon. For instance, Huber, Franz, and Bülthoff (1999) demonstrate a minimalistic model of the visual orientation and fixation behavior of flies on a mobile robot equipped with a catadioptric imaging system and a conical mirror. The same robot is used by Franz et al. (1998b) to implement an insect-inspired strategy for visual homing. Both studies use one-dimensional, panoramic views of the environment. Franz et al. (1999) employ insect-inspired matched filters to estimate the self-motion of a mobile robot. They extend the vertical field of view to a narrow band around the horizon in order to apply a two-dimensional, Cartesian grid of EMDs to the panoramic image.

A full spherical field of view is used in a recent simulation study by Tammero and Dickinson (2002) to reconstruct the optic flow experienced by a fruit fly during free flight. Here, all image processing is homogeneously performed in the Mercator plane. As explained in Section 3.1.2, this results in space-variant filtering of the actual spherical image. In particular, the base distances of the EMDs located between adjacent pixels vary with the elevation angle. However, in this particular study the distortions are not problematic since the visual stimulus presented to both the original fly and to the simulation model is restricted to a textured, panoramic cylinder. Both the ground plane and the ceiling are uniformly black. In addition, the simulated visual system moves mainly in horizontal directions and has a fixed orientation with respect to the environment, allowing only yaw rotations. This ensures that in most situations the stimulus covers only a narrow, horizontal band in the spherical field of view where the distortions are less pronounced than for larger elevation angles.

In general, however, panoramic image processing is not sufficient for a comprehensive model of the insect visual system. Flying insects have more degrees of freedom than ground-based robots, and they may change their altitude or rotate about arbitrary axes, thereby experiencing a much wider variety of self-motion-induced, spherical optic flow fields. In addition, there is evidence that insects use ventral optic flow induced by ground motion for visual altitude control (Srinivasan et al., 2000). Thus, modeling the insect visual system to study visual flight control requires a full spherical field of view.

An accurate and efficient simulation model for insect compound eyes was presented in Chapter 2. In this chapter I introduce a novel method for the efficient and homogeneous representation and processing of spherical images. It closely approximates insect vision and eliminates the disadvantages of traditional, planar representations.

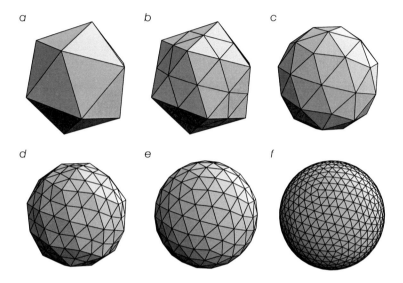

Figure 3.3: Construction of a spherical geodesic grid. The edges of an icosahedron (*a*) are bisected (*b*), and the resulting new vertices are projected onto the unit sphere (*c*). Repeated subdivisions (*d*) increase the spatial resolution of the spherical grid (*e*, *f*).

3.2 Geodesic Grids

Geodesic grids are quasi-uniform tesselations of the sphere which are generated by recursive subdivision of a Platonic solid. They were invented in the 1920s by Walter Bauersfeld of the Carl Zeiss Optical Company in Jena, Germany for the construction of planetarium projection domes. 30 years later the term 'geodesic dome' was introduced by R. Buckminster Fuller who popularized polyhedral buildings in the United States. In the late 1960s geodesic grids were suggested for the representation of spherical data in the context of atmospheric modeling (Sadourny, Arakawa, and Mintz, 1968; Williamson, 1968). In recent years they received renewed attention in global climate simulations (Heikes and Randall, 1995; Randall, Ringler, Heikes, Jones, and Baumgardner, 2002).

In this study I propose geodesic grids as a convenient data structure for the efficient representation and processing of discrete spherical images.

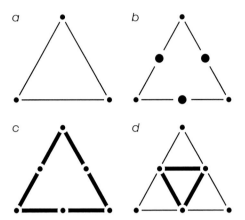

Figure 3.4: Subdivision of an individual face of an icosahedron. Each edge of the original triangle (*a*) is bisected by a newly inserted vertex (*b*) and replaced by two new edges (*c*). Each original triangle is subdivided into four new triangles by three additional edges (*d*).

3.2.1   Construction

The construction of a spherical geodesic grid is illustrated in Fig. 3.3. The starting point is an icosahedron inscribed to the unit sphere (Fig. 3.3*a*) for which we define a resolution level $R = 0$. To generate the first level of subdivision, each edge of the original icosahedron is bisected by a newly inserted vertex, and replaced by two new edges. Each triangular face is subdivided into four smaller triangles by three additional edges connecting the new vertices (Fig. 3.3*b*). All new vertices are projected onto the unit sphere to yield a geodesic grid of resolution level $R = 1$ (Fig. 3.3*c*). This subdivision procedure can be applied repeatedly (Fig. 3.3*d*) to increase the spatial resolution of the geodesic grid to an arbitrary level (Fig. 3.3*e,f*). A detailed subdivision scheme for an individual, triangular face is depicted in Fig. 3.4.

3.2.2   Properties

An icosahedron-based geodesic grid of resolution level R comprises a set $F^{(R)}$ of faces with

$$|F^{(R)}| = 20 \cdot 4^R, \qquad (3.1)$$

a set $E^{(R)}$ of edges with

$$|E^{(R)}| = 30 \cdot 4^R, \qquad (3.2)$$

R	$|V^{(R)}|$	$|E^{(R)}|$	$|F^{(R)}|$
0	12	30	20
1	42	120	80
2	162	480	320
3	642	1920	1280
4	2562	7680	5120
5	10242	30720	20480
6	40962	122880	81920

Table 3.1: Number of vertices, edges and faces of an icosahedron-based spherical geodesic grid. R is the level of subdivision and determines the resolution of the grid, starting with $R = 0$ for the original icosahedron.

and a set $V^{(R)}$ of vertices. After Euler's well-known theorem the number of vertices is

$$|V^{(R)}| \;=\; |E^{(R)}| - |F^{(R)}| + 2 \tag{3.3}$$
$$=\; 10 \cdot 4^R + 2 \,. \tag{3.4}$$

Tab. 3.1 gives an overview of the numbers of vertices, edges and faces for the first seven levels of resolution.

Almost all vertices of an icosahedron-based geodesic grid have six neighbors. The only exceptions are the 12 vertices of the original icosahedron which have five neighbors and exist in all resolution levels. Thus, the surface tiling of the sphere induced by the vertices of a geodesic grid always contains twelve pentagonal tiles, whereas all other tiles are hexagonal.

Hexagons exhibit a higher degree of symmetry and isotropy than the other two regular polygons completely tiling the plane, i.e., triangles and squares. As illustrated in Fig. 3.5, each tile in a triangular or rectangular grid has at least two different types of neighbors. Edge neighbors share one edge and two vertices, whereas vertex neighbors share only one common vertex. Using this definition, each triangular tile has three edge neighbors and nine vertex neighbors, whereas each square tile has four edge neighbors and four vertex neighbors. Different types of local neighborhoods might require different treatment in the processing of information associated with each vertex, e.g., due to different distances between adjacent vertices. In contrast, icosahedral geodesic grids do not contain any vertex neighbors. All hexagonal (Fig. 3.5c) and pentagonal (Fig. 3.5f) tiles have only edge neighbors with an approximately equal distance from the center tile. Thus, local neighborhoods on an icosahedral geodesic grid are isotropic, and all information stored on adjacent vertices can be treated equally.

In addition, geodesic grids provide a quasi-homogeneous distribution of vertices over the entire sphere. One possible measure for this homogeneity is the angular distance between adjacent vertices. Tab. 3.2 shows a relative standard deviation from

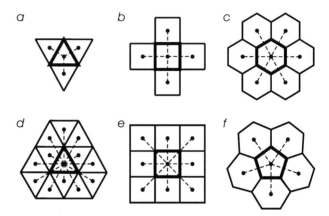

Figure 3.5: Symmetry and isotropy of different surface tilings. Triangular (*a, d*) and square (*b, e*) grids exhibit at least two classes of neighborhoods with different distances between the tile centers. Triangular tiles have three edge neighbors (*a*) and nine additional vertex neighbors (*d*). Square tiles have four edge neighbors (*b*) and four additional vertex neighbors (*e*). In contrast, hexagonal tiles have only edge neighbors of equal distance (*c*) and therefore exhibit the highest degree of symmetry. Both the hexagonal and the 12 pentagonal (*f*) tiles induced by the vertices of an icosahedron-based geodesic grid are surrounded exclusively by edge neighbors. Icosahedral grids do not contain any vertex neighbors.

the mean inter-vertex angle of $\sigma/\varphi_{\mathrm{mean}} = 0.065$, and a ratio of minimum to maximum angle of $\varphi_{\min}/\varphi_{\max} = 0.84$, both indicating that the inter-vertex angle is almost constant within each level of resolution. The slight variations are due to the grid generation by recursive subdivision and could be further reduced by iterative optimization methods. The inter-vertex angle is of particular importance for the implementation of correlation-type EMDs described in Section 3.6.2.

3.2.3   Image Representation

The most common method to store and process an image in computer memory is to sample its intensity or color distribution at discrete spatial locations (Jähne, 2001). Planar images are usually represented on a Cartesian sampling grid. However, as shown in Section 3.1.2, rectangular grids are not suitable for spherical images. This study proposes to represent spherical images on geodesic grids using the vertex positions of a subdivided icosahedron as sampling points. The black dots in Fig. 3.6 indicate the location of sampling points in a spherical field of view for the first four resolution levels

R	$\varphi_{\text{mean}} \, [^\circ]$	$\sigma \, [^\circ]$	$\sigma/\varphi_{\text{mean}}$	$\varphi_{\text{min}} \, [^\circ]$	$\varphi_{\text{max}} \, [^\circ]$	$\varphi_{\text{min}}/\varphi_{\text{max}}$
0	63.43	0	0	63.43	63.43	1
1	33.86	2.15	0.064	31.72	36.00	0.88
2	17.22	1.12	0.065	15.86	18.70	0.85
3	8.64	0.56	0.065	7.93	9.44	0.84
4	4.33	0.28	0.065	3.96	4.73	0.84
5	2.16	0.14	0.065	1.98	2.37	0.84
6	1.08	0.07	0.065	0.99	1.18	0.84

Table 3.2: Angles between adjacent vertices of a spherical geodesic grid. R is the level of resolution, φ_{mean} is the mean inter-vertex angle in degrees. σ and $\sigma/\varphi_{\text{mean}}$ are the absolute and relative standard deviations, respectively. φ_{min}, φ_{max} and $\varphi_{\text{min}}/\varphi_{\text{max}}$ are the minimum and maximum angles and the ratio of minimum to maximum angle, respectively.

$R \in \{0, \ldots, 3\}$ of an icosahedral geodesic grid. The color or gray value information is stored on the vertices of the geodesic grid. Each vertex corresponds to one pixel.

To visualize a spherical image, a portion of the surface area needs to be assigned to each pixel since the vertices on which the color information resides are dimensionless points. This is achieved by constructing a second geodesic grid dual to the original one, i.e., each face of the dual grid corresponds to one vertex of the original subdivided icosahedron. In this study the centers of gravity of the triangular faces of the original grid are used as the vertices of the dual grid. The black lines in Fig. 3.6 indicate the resulting pixel boundaries. Note that the pixels may have different shapes and sizes when alternative partitions such as spherical Voronoi meshes (Augenbaum and Peskin, 1985) are used. However, the particular choice of pixel boundaries affects only the visualization of an image, but not the image itself since the vertex positions of the original subdivided icosahedron (on which the image information is stored) remain unchanged.

This study further proposes geodesic grids as an innovative method for representing directional information such as optic flow fields on the sphere. In contrast to conventional vector field representations, the optic flow is decomposed into one-dimensional, local components defined by the edges of a geodesic grid. The thick black lines in Fig. 3.7 indicate the edges of icosahedral geodesic grids of the first four resolution levels $R \in \{0, \ldots, 3\}$. Each edge has a fixed position and orientation on the sphere and can be used to represent one local, one-dimensional component of a spherical flow field as a scalar value. Geodesic grids are therefore especially useful for implementing spherical arrays of insect-inspired, correlation-type elementary motion detectors. Section 3.6.2 provides a detailed description of local motion detection on geodesic grids.

Fig. 3.8 shows examples of discrete spherical images on geodesic grids. The im-

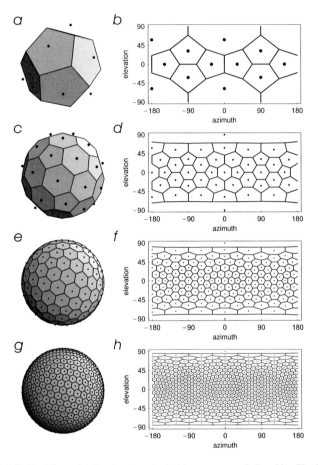

Figure 3.6: Pixel formation for discrete spherical images on geodesic grids. The first four resolution levels of an icosahedral grid are shown as 3D models (*left column*) and as Mercator projections (*right column*). Black dots indicate the vertices of a subdivided icosahedron which define the sampling directions of the spherical image. The pixel boundaries are depicted as black lines. The corners of each pixel are formed by the centers of gravity of the triangular faces of the subdivided icosahedron.

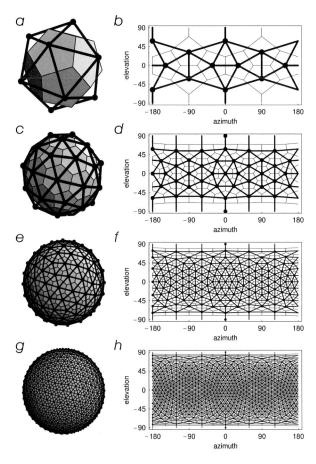

Figure 3.7: Representation of discrete spherical images on geodesic grids. The first four resolution levels of an icosahedral grid are depicted as 3D models (*left column*) and as Mercator projections (*right column*). The vertices (black dots) serve as receptor viewing directions, whereas the edges (thick lines) represent elementary motion detectors connecting adjacent receptor units. Thus, color or gray value information is stored on the vertices, whereas motion information resides on the edges of the grid. Thin lines indicate pixel boundaries.

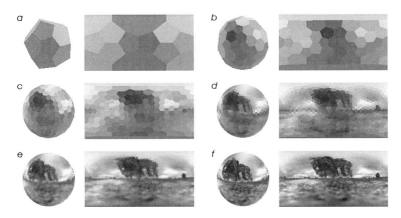

Figure 3.8: Spherical images on geodesic grids. The example scene is shown for six differ-ent resolution levels $R \in \{0, \ldots, 5\}$ in pairs of 3D spherical (*left*) and 2D Mercator (*right*) projections. The images were acquired in a highly realistic 3D virtual environment using the omnidirectional compound eye simulation model described in Chapter 2.

ages were acquired for six different levels of resolution $R \in \{0, \ldots, 5\}$ in a highly realistic 3D virtual environment, using the omnidirectional compound eye model de-scribed in Chapter 2. Each image is shown as a 3D sphere and as a 2D Mercator projection.

In conclusion, the vertices and edges of icosahedral geodesic grids are distributed over the sphere in a quasi-homogeneous and isotropic fashion. Geodesic grids there-fore provide an excellent sampling grid for spherical images representing intensity or local motion. The following sections introduce a graph-based notation and describe various algorithms for spherical image processing on geodesic grids.

3.3 Discrete Spherical Image Processing

3.3.1 Geodesic Grids as Graphs

For image processing it is convenient to consider a spherical geodesic grid of resolution level $R \in \{0, 1, 2, \ldots\}$ as a graph

$$G^{(R)} = (V^{(R)}, E^{(R)}, v_1^{(R)}, v_2^{(R)}) \tag{3.5}$$

with a set $V^{(R)}$ of vertex indices

$$v \in V^{(R)} = \{1, 2, \ldots, |V^{(R)}|\} \tag{3.6}$$

and a set $E^{(R)}$ of edge indices

$$e \in E^{(R)} = \{1, 2, \ldots, |E^{(R)}|\} . \tag{3.7}$$

The connectivity of the graph is described by two functions

$$v_1^{(R)}, v_2^{(R)} : E^{(R)} \to V^{(R)} \tag{3.8}$$

which return the indices of the first and second vertex of each edge, respectively. For instance, $v_1^{(R)}(3)$ yields the index of the first vertex of the third edge. This notation facilitates the description of image representation and processing following below. The position of each vertex on the unit sphere is defined by the function

$$\mathbf{v}^{(R)} : V^{(R)} \to U = \{\mathbf{u} \in \Re^3 \mid \mathbf{u} \cdot \mathbf{u} = 1\} . \tag{3.9}$$

As stated above, the color or gray value information of a spherical image resides on the vertices of the geodesic grid. Thus, the vector

$$I^{(R)} \in \Re^{|V^{(R)}|} \tag{3.10}$$

contains a complete spherical gray value image of resolution level R. An extension for color images is straightforward. Information on local, one-dimensional components of image motion is stored on the edges of the geodesic grid using the vector

$$\Phi^{(R)} \in \Re^{|E^{(R)}|} . \tag{3.11}$$

3.3.2 Linear Filtering on the Sphere

In the most general case, each pixel in the spherical target image $I_T^{(R)}$ depends on all pixels in the source image $I_S^{(R)}$, thus filtering is performed by the following algorithm:

```
for all vertices  v ∈ {1,...,|V^(R)|}
    I_T^(R)[v]  :=  ∑_{i=1}^{|V^(R)|} K(v^(R)(i), v^(R)(v)) · I_S^(R)[i]
```

The spatial relationship between pixels is determined directly from their positions on the unit sphere, and the filter mask may have an arbitrary size and shape. The filter kernel is defined by a function

$$K(\mathbf{u}_S, \mathbf{u}_T) : U \times U \to \Re \tag{3.12}$$

which is centered at the target direction $\mathbf{u}_T$ and returns the local weight for each source direction $\mathbf{u}_S$. For instance, an isotropic Gaussian with maximum at $\mathbf{u}_T$ can be defined as a function of the angle between $\mathbf{u}_T$ and $\mathbf{u}_S$. In general, this permits the definition of arbitrary and even space-variant linear filters, but the algorithm requires $O(|V^{(R)}|^2)$ operations. In practice, however, the sensitivity of most filter kernels such as Gaussians is localized around the target direction $\mathbf{u}_T$ and zero or close to zero for all other regions of the sphere. Thus, processing time can be strongly reduced by using a lookup table storing only non-zero weights together with the corresponding pixel indices.

3.3.3   Graph-Based Spherical Image Processing

All image processing operations described in the following sections are based on the graph structure of geodesic grids. The spatial configuration of pixels is determined from the edge information stored in the functions $v_1^{(R)}(e)$ and $v_2^{(R)}(e)$. Here, only directly adjacent pixels are used as input values for local operations. This results in filter masks with a diameter of three pixels, consisting of a hexagonal center and six surrounding pixels, or a pentagonal center and five surrounding pixels, respectively. Thus, each filter mask exists in two versions, one for hexagonal and one for pentagonal pixels. An isotropic filter therefore requires four weighting factors w_{c6}, w_{s6}, w_{c5} and w_{s5} for hexagon center, hexagon surround, pentagon center and pentagon surround pixels, respectively. The weighting factors are illustrated in Fig. 3.9*a* and *d*.

Vertex Sorting

When processing an entire spherical image, each pixel needs to be identified as a hexagon or pentagon in order to apply the correct hexagonal or pentagonal filter mask. Sorting the entries of the vertex and edge lists in an appropriate manner allows to derive this information directly from the index of the currently processed vertex or edge. Both lists need to be sorted only once for each resolution level before the filtering operations are performed. This makes geodesic grid-based spherical image processing very efficient since time-consuming search operations or complex address calculations are completely avoided.

The construction method for icosahedron-based geodesic grids implies that a grid of resolution $R + 1$ includes all vertices of resolution R. Thus, due to recursion, a grid of resolution R contains all vertices of all $R + 1$ resolution levels in $\{0, 1, \ldots, R\}$. Each vertex can therefore be assigned to one of $R + 1$ distinct groups according to the resolution level in which it was generated. If these groups are arranged in ascending order, their boundaries can be determined using Eq. 3.4. In particular, the 12 pentagonal pixels inherited from the original icosahedron always occupy the first 12 indices of the vertex list. Thus, sorting the vertices allows to determine the resolution level in which a specific vertex was generated, and provides a simple test whether a specific pixel has pentagonal or hexagonal shape.

The computation of spherical resolution pyramids requires additional sorting of the vertices. Each vertex of resolution $R + 1$ that already exists in resolution R must have the same index in both resolutions. This aligns the low and high resolution spherical images and thereby facilitates the exchange of image information between adjacent resolution levels. A sorting of the vertices that meets all of these conditions is automatically obtained when a grid of resolution $R + 1$ is constructed by first copying the entire, unmodified vertex list of resolution R into the new vertex list, and subsequently appending all newly generated vertices to the end of the new list.

Edge Sorting

Applying a filter mask to a specific pixel requires the color or gray value information from all directly surrounding pixels. The image processing operations discussed below use the edge list of the geodesic grid to determine adjacent pixels. However, the edges connecting a specific pixel with its immediate neighbors are distributed over the entire edge list and do not have consecutive indices. The edge list cannot be sorted by pixel indices since the two pixels connected by each edge mutually serve as center and surround pixels of the according filter masks. Thus, finding all neighbors of a specific pixel requires searching the entire edge list for edges originating from that pixel. However, this search needs not be performed separately for each target pixel. The graph-based filtering method proposed below utilizes the fact that the edge list contains all neighbors of all vertices in the geodesic grid, thus a single pass through the edge list is sufficient to apply the filter mask simultaneously to all pixels of the spherical image.

To select the correct pentagonal or hexagonal filter mask both pixels connected by each edge need to be identified as having five or six neighbors, respectively. Testing each pixel individually, e.g., using the sorted vertex list described above and testing for a vertex index $v \in \{1, \dots, 12\}$ to identify pentagons, would require two tests per edge. The algorithm proposed here avoids these tests by sorting the edge list so that all 60 edges connected to one of the 12 pentagonal pixels are represented by the first 60 indices. The number of 60 pentagon-hexagon edges remains constant for all resolution levels $R \geq 1$. The only exception is the original icosahedron ($R = 0$) which contains 30 pentagon-pentagon edges.

Additional sorting of the edge list is required for spherical resolution pyramids. Both algorithms for image reduction and image expansion described below use only such edges of the high resolution grid ($R + 1$) which connect a vertex newly generated in resolution $R + 1$ with a vertex inherited from the lower resolution level R. When a geodesic grid of resolution $R + 1$ is generated from a grid of resolution R, each of the

	Memory [kB]	
R	Single	Pyramid
0	0.23	0.23
1	0.94	1.17
2	3.75	4.92
3	15.50	19.92
4	60.00	79.92
5	240.00	319.92
6	960.00	1279.92

Table 3.3: Memory requirements for the edge list of icosahedral geodesic grids. The vertex indices stored in the edge list are assumed to be 4-byte integer numbers. Note that the edge lists of all resolution levels together (*right column*) require only an additional $1/3$ of the memory used by the highest level of resolution alone (*left column*).

$|E^{(R)}|$ edges in the lower resolution level is bisected and replaced by two new edges, so that each of the resulting $2|E^{(R)}|$ edges is connected to an old and a new vertex. Using Eq. 3.2, the number of edges of this type is

$$2|E^{(R)}| = \frac{1}{2}|E^{(R+1)}|\,, \tag{3.13}$$

i.e., half of the edges of resolution level $R + 1$ connect old and new vertices. All remaining edges connect two new vertices and are not used for image reduction or expansion. Thus, grouping all edges of the first type at the beginning of the edge list provides immediate access to all relevant edges and facilitates the computation of spherical resolution pyramids.

Edge List Size

The edge list of a geodesic grid contains two vertex indices for each edge and defines all pixel neighborhoods. It needs to be stored only once for each resolution level and is used by all local image processing operations involving interactions with adjacent pixels. According to Eq. 3.2 the number of edges in a single resolution level R is $30 \cdot 4^R$. The ratio of the sum of all edges in all resolution levels $i \in \{0, \ldots, R - 1\}$ to the number of edges in the highest level R is

$$\lim_{R \to \infty} \left(\frac{\sum_{i=0}^{R-1} 30 \cdot 4^i}{30 \cdot 4^R} \right) = \lim_{R \to \infty} \left(\frac{1}{3} - \frac{1}{3 \cdot 4^R} \right) = \frac{1}{3}\,, \tag{3.14}$$

thus the edge lists of all resolution levels together require only an additional $1/3$ of the memory used by the highest level of resolution alone. Tab. 3.3 shows the memory requirements for the first seven levels of resolution, assuming 4-byte integer numbers as vertex indices. Note that for resolution levels $R \leq 6$, i.e., for a maximum number of $|V^{(6)}| = 40962$ pixels, 2-byte unsigned integer numbers are sufficient to represent all vertex indices and can be used to further reduce the size of the lookup table by a factor of $1/2$. The highest resolution observed in insect compound eyes (cf. Chapter 2) corresponds roughly to a grid of resolution $R = 6$.

3.4 Graph-Based Spherical Convolution

One of the most elementary image processing operations is the convolution of a complete image with an isotropic kernel. Usually the kernel is a Gaussian or a Difference of Gaussians (DOG) function and has a small diameter compared to the image size. For planar images with rectangular pixels, square filter masks with a size of three by three pixels are commonly used to approximate the convolution kernel. In analogy, spherical convolution can be applied to geodesic grid-based images using hexagonal and pentagonal filter masks with a diameter of three pixels.

The convolution with a Gaussian kernel smooths an image by suppressing high spatial frequencies. Local intensity differences between each pixel and its neighbors are removed by local weighted averaging. The resulting spatial low pass filtering of the image is useful for noise reduction since in most practical situations the amount of noise increases with spatial frequency, e.g., due to uncorrelated noise of the individual pixels of a camera or in the light receptors of a retina. Fig. 3.9*b* and *e* show low pass filter masks for hexagonal and pentagonal pixels. Both masks are isotropic, contain positive weights for both center and surrounding pixels, and have a unit gain factor.

In contrast, convolution with a DOG kernel results in spatial band pass filtering. A DOG function is composed of a narrow, positive Gaussian in the center and a wider, negative Gaussian in the surrounding area which define the upper and lower spatial frequencies (cutoff frequencies) of the filter transmission band (Jähne, 2001). Here, a filter mask with a diameter of three pixels subtracts the average intensity of the surrounding pixels from the center pixel. In biological information processing this operation is known as lateral inhibition. It removes the average local background signal or DC component from each pixel and thereby enhances the contrast of the image. Since the image is composed of discrete pixels there is no further spatial resolution within the center pixel of the filter mask, the upper cutoff frequency of the band pass filter is identical with the highest possible spatial frequency which can be represented in the image. Therefore, in this case the band pass filter is equivalent to a high pass. Isotropic hexagonal and pentagonal high pass filter masks are depicted in Fig. 3.9*c* and *f*. The center pixel of each filter mask is positive, whereas all surrounding pixels have negative weights. Both the positive and the negative components have a unit gain factor.

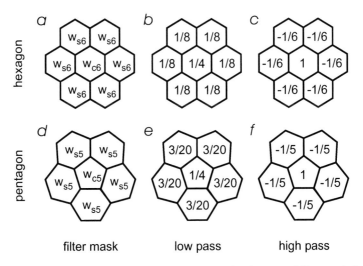

Figure 3.9: Isotropic filter kernels for hexagonal (*upper row*) and pentagonal (*lower row*) pixel neighborhoods. *a,d.* The weighting factors for center and surround pixels of hexagonal and pentagonal filter masks are w_{c6}, w_{s6}, w_{c5} and w_{s5}, respectively. *b,e.* Isotropic low pass filter for noise reduction and image smoothing by local averaging. *c,f.* Isotropic high pass filter for contrast enhancement by lateral inhibition.

Algorithm

Applying a filter mask to each pixel in a spherical source image $I_S^{(R)}$ of resolution R in order to obtain a target image $I_T^{(R)}$ of equal resolution requires two successive processing steps. In the first step each source pixel serves as the center pixel of the filter mask and is copied into the corresponding target pixel after the appropriate weighting factor is applied. This requires one pass through the entire vertex list. Since the vertex list is sorted by the resolution level in which each vertex is generated, the 12 pentagonal pixels are located at the beginning of the vertex list. They are weighted as pentagon centers using w_{c5}, whereas all other pixels are weighted as hexagon centers using w_{c6}.

```
for all vertices v ∈ {1,...,12}
    I_T^(R)[v]  := w_c5 · I_S^(R)[v]
for all vertices v ∈ {13,...,|V^(R)|}
    I_T^(R)[v]  := w_c6 · I_S^(R)[v]
```

In the second processing step the source pixels are weighted as the surrounding pixels of the filter mask, and added to the corresponding center pixels in the target image. This requires one pass through the entire edge list. For $R \geq 1$ the sorting of the edge list ensures that each of the first 60 edges is connected to one of the 12 pentagons and to one hexagon, whereas each of the remaining edges connects two hexagons. Thus, one vertex of each of the first 60 edges is weighted with the pentagon surround weight w_{s5}, the other vertex is weighted with the hexagon surround weight w_{s6}. For all other edges, both vertices are weighted as hexagon surround pixels using w_{s6}.

```
for all edges e ∈ {1,...,60} {
    I_T^(R)[v_1^(R)(e)]  :=  I_T^(R)[v_1^(R)(e)] + w_s5 · I_S^(R)[v_2^(R)(e)]
    I_T^(R)[v_2^(R)(e)]  :=  I_T^(R)[v_2^(R)(e)] + w_s6 · I_S^(R)[v_1^(R)(e)]
}
for all edges e ∈ {61,...,|E^(R)|} {
    I_T^(R)[v_1^(R)(e)]  :=  I_T^(R)[v_1^(R)(e)] + w_s6 · I_S^(R)[v_2^(R)(e)]
    I_T^(R)[v_2^(R)(e)]  :=  I_T^(R)[v_2^(R)(e)] + w_s6 · I_S^(R)[v_1^(R)(e)]
}
```

Note that this implementation does not require any tests within the processing loop to identify pentagonal pixels. However, the original icosahedron ($R = 0$) is a special case since all of its 12 vertices have five neighbors. Therefore, the second processing step applies the pentagon surround weight w_{s5} to both vertices of each edge.

```
for all edges e ∈ {1,...,30} {
    I_T^(R)[v_1^(R)(e)]  :=  I_T^(R)[v_1^(R)(e)] + w_s5 · I_S^(R)[v_2^(R)(e)]
    I_T^(R)[v_2^(R)(e)]  :=  I_T^(R)[v_2^(R)(e)] + w_s5 · I_S^(R)[v_1^(R)(e)]
}
```

3.5 Spherical Resolution Pyramids

A resolution pyramid is a set of images $\{I^{(0)}, \ldots, I^{(R)}\}$ with different spatial resolutions which are all generated from a single original image $I^{(R)}$. In planar image processing, resolution pyramids are a well-known method to efficiently implement filtering operators with a large spatial extent (Jähne, 2001). Instead of processing a large number of pixels in the original, high resolution image, a small filter mask is sufficient to perform the same operation in a reduced version of the image. Resolution pyramids have a multitude of applications, for instance the so-called MIP mapping used in computer graphics to sample textures with filter masks of constant size (Williams, 1983). This section shows how resolution pyramids are computed for geodesic grid-based spherical images.

In planar resolution pyramids the side length of an image $I^{(R)}$ is usually reduced by a factor of $1/2$ in the next lower resolution level, thus the number of pixels in the low resolution image $I^{(R-1)}$ is $1/4$ of the original image. This makes both the computation of resolution pyramids and their storage in computer memory highly efficient. An entire pyramid including all resolution levels requires only an additional $1/3$ of the memory used by the highest level of resolution alone (Jähne, 2001). This also holds for geodesic grid-based spherical resolution pyramids. According to Eq. 3.4, a single spherical image of resolution R contains $10 \cdot 4^R + 2$ pixels. For resolutions $R \geq 1$ the ratio of the sum of all lower resolution levels $i \in \{0, \dots, R-1\}$ to the number of pixels in the highest level is

$$\lim_{R \to \infty} \left(\frac{\sum_{i=0}^{R-1}(10 \cdot 4^i + 2)}{10 \cdot 4^R + 2} \right) = \lim_{R \to \infty} \left(\frac{1}{3} + \frac{R-2}{5 \cdot 4^R + 1} \right) = \frac{1}{3}. \qquad (3.15)$$

Thus, only $1/3$ of the number of pixels of the highest resolving spherical image need to be processed and stored for all other resolution levels together.

Gaussian and Laplacian Pyramids

The most commonly used planar resolution pyramids are the Gaussian and the Laplacian pyramids. A Gaussian pyramid is constructed by successively reducing the resolution of an image, usually by sub-sampling with twice the original sampling distance. However, only spatial frequencies below the Nyquist frequency of the sampling grid are correctly represented in the image, i.e., spatial frequencies sampled at least twice per cycle (Mallot, 2000; Jähne, 2001). Higher spatial frequencies would lead to spatial aliasing and need to be suppressed by a spatial low pass filter averaging over the pixels surrounding each sampling point. Thus, the maximum spatial frequency in each resolution level of a Gaussian pyramid is reduced by a factor of $1/2$ compared to the next higher level. A reduced image can be expanded to its original size by super-sampling, using the original, smaller sampling distance. Naturally, high spatial frequencies eliminated during image reduction cannot be reconstructed, thus the color or gray values of high resolution pixels located between the low resolution sampling points need to be determined by interpolation. The resulting image is a low pass filtered version of the original image and approximates a convolution with a Gaussian kernel.

A Laplacian pyramid (Burt and Adelson, 1983) contains the complete information of the original image in a set of spatially band pass filtered images with successively lower resolutions. Each image represents all spatial frequencies in the interval between its own Nyquist frequency and that of the next lower resolution level. Computing Laplacian pyramids is very efficient. Each level of resolution can be determined by subtracting two consecutive resolution levels of the corresponding Gaussian pyramid. In particular, the image with lower resolution in the Gaussian pyramid is expanded and subsequently subtracted from the higher resolution image, thereby approximating a convolution with a DOG kernel. The original image can be reconstructed without loss of information by successive expansion and addition of all images in the Laplacian pyramid.

Spherical Resolution Pyramids on Geodesic Grids

The recursive structure of geodesic grids makes them especially useful for the implementation of spherical resolution pyramids. In analogy to planar resolution pyramids this requires two image processing operations: one to reduce an image from resolution level $R + 1$ to level R, and one to expand it from level R to level $R + 1$. The pixels associated with the vertices of the lower and higher resolution grids are referred to as major and minor pixels, respectively.

As in the planar case, the geodesic grid-based reduction of a spherical image is equivalent to the convolution with a Gaussian low pass kernel and the subsequent subsampling onto a grid of lower resolution. Low pass filtering is required to prevent spatial aliasing, i.e., artifacts due to high spatial frequencies in the original image which cannot be represented on the low resolution grid. Repeated reduction of an image yields a spherical Gaussian pyramid.

Spherical image expansion increases the spatial resolution by super-sampling the image onto a finer grid. The color or gray values of all pixels with coincident sampling directions in both the low and high resolution images can be copied directly. The values of all other pixels in the high resolution image need to be determined by interpolation. Expanding a previously reduced image to its original resolution results in a spatially low pass filtered version of the original image and approximates a convolution with a Gaussian kernel. As in the planar case, subtracting the expanded image from the original approximates a convolution with a DOG kernel and yields a spatially band pass filtered version of the original image. This can be applied to all resolution levels of a spherical Gaussian pyramid and provides an efficient method to generate a spherical Laplacian pyramid. As in the planar case, successive expansion and addition of all images in the Laplacian pyramid allows to reconstruct the original high resolution spherical image without loss of information and independently of the chosen interpolation method.

3.5.1 Image Reduction

A simple and fast method to reduce planar images with rectangular pixels to a quarter of their original size is to average over square blocks of two by two pixels. For spherical images on geodesic grids the situation is more complex since hexagonal pixels cannot be aggregated to larger hexagons or tiled by smaller hexagons. In addition, special

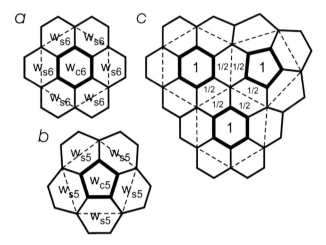

Figure 3.10: Image reduction and expansion between consecutive levels of resolution on a spherical icosahedral grid. Dashed lines indicate the pixel boundaries of the low resolution image (major pixels), whereas pixels of the next higher resolution level are enclosed by solid lines (minor pixels). *a,b.* Image reduction for hexagonal (*a*) and pentagonal (*b*) pixels. The gray or color values of the low resolution target image are determined by sub-sampling the high resolution source image. An isotropic low pass filter is used to average over the center pixel (thick solid lines) and all adjacent minor pixels. *c.* Image expansion for hexagonal (*upper left, lower center*) and pentagonal (*upper right*) pixels. The gray values of the low resolution source image are copied to the center pixels (thick solid lines), and interpolated for all other pixels of the high resolution target image.

treatment is required for the 12 pentagons contained in each geodesic grid.

   The resolution of a spherical image is reduced by sub-sampling the original image at the vertices of a geodesic grid of lower resolution. Spatial aliasing is inhibited by overlapping low pass filter masks as shown in Fig. 3.10*a* and *b*. The filter masks must have the following properties:

1. The filter mask must be isotropic. A mask with a diameter of three pixels may contain only two different weights w_c and w_s for the center and for the surrounding pixels, respectively. As in the previous section, pentagonal filter kernels use w_{c5} and w_{s5}, whereas hexagonal kernels use w_{c6} and w_{s6}.

2. The filter mask must have a unit gain factor, i.e., the sum of all of its weights must equal 1.

3. All pixels in the high resolution source image must contribute equally to the low resolution target image. Since the surrounding source pixels contribute to two overlapping filter masks they receive half the weight of the center pixels.

Solving the corresponding equation system for hexagonal pixels yields the filter mask shown in Fig. 3.9*b*.

However, not all of these conditions can be fulfilled simultaneously for both hexagonal and pentagonal filter masks since they overlap at all pixels surrounding one of the 12 pentagons. For $w_{s6} \neq w_{s5}$, condition (3) is violated since the contribution $w_{s6} + w_{s5}$ of the hexagon-pentagon overlap pixels to the target image does not equal the hexagon-hexagon overlap contribution $2w_{s6}$ or the pentagon-pentagon overlap contribution $2w_{s5}$. For $w_{s6} = w_{s5}$, one the center weights w_{c6} or w_{c5} needs to be adapted in order to fulfill condition (2). Again, this violates condition (3) since then $w_{c6} \neq w_{c5}$. Here, a pentagonal kernel is chosen which fulfills conditions (1) and (2) but not condition (3). It is depicted in Fig. 3.9*e*. The error $w_{s5} - w_{s6} = 1/40$ introduced by relaxing condition (3) is very small and occurs only with the 12 pentagonal pixels. It does not noticeably affect subsequent processing stages such as local motion detection or large field integration.

Algorithm

Reducing a spherical source image $I_S^{(R+1)}$ of resolution $R + 1$ to obtain a target image $I_T^{(R)}$ of lower resolution R resembles the graph-based convolution described in the previous section. It includes low pass filtering of the high resolution source image and requires one pass through the vertex list and one pass through the edge list. In contrast to convolution, the source image is sub-sampled at the vertex positions of the low resolution grid. By construction, each sampling point of the low resolution image coincides with one vertex of the high resolution grid. The corresponding pixels of the high resolution image are therefore weighted as the center pixels of the low pass filter mask, using w_{c5} for pentagonal and w_{c6} for hexagonal pixels, respectively. The vertex sorting ensures that the vertices of the low resolution grid occupy the first $|V^{(R)}|$ indices of the high resolution vertex list. The first 12 vertices represent pentagonal pixels, whereas all remaining pixels are hexagonal.

```
for all vertices v ∈ {1, ..., 12}
    I_T^(R)[v]  :=  w_c5 · I_S^(R+1)[v]

for all vertices v ∈ {13, ..., |V^(R)|}
    I_T^(R)[v]  :=  w_c6 · I_S^(R+1)[v]
```

All pixels of the high resolution grid located between the sampling directions of the low resolution target image are weighted as surrounding pixels of the low pass filter mask. The edge list of the high resolution grid is used to assign these pixels to their corresponding center pixels. In contrast to the convolution algorithm described in the previous section, each edge is used only in one direction since only one of its vertices represents a target pixel. In addition, only the first $\frac{1}{2}|E^{(R+1)}|$ edges of the high resolution grid are connected to a target vertex. Due to edge sorting, each of the first 60 edges connects a pentagonal center with a hexagonal surrounding pixel which is therefore weighted with w_{s5}. All remaining edges originate from hexagonal center pixels, thus the corresponding surrounding pixels are weighted with w_{s6}.

$$
\begin{aligned}
&\texttt{for all edges } e \in \{1, \ldots, 60\} \\
&\quad I_T^{(R)}[v_1^{(R+1)}(e)] \ := \ I_T^{(R)}[v_1^{(R+1)}(e)] + w_{s5} \cdot I_S^{(R+1)}[v_2^{(R+1)}(e)]
\end{aligned}
$$

$$
\begin{aligned}
&\texttt{for all edges } e \in \{61, \ldots, \tfrac{1}{2}|E^{(R+1)}|\} \\
&\quad I_T^{(R)}[v_1^{(R+1)}(e)] \ := \ I_T^{(R)}[v_1^{(R+1)}(e)] + w_{s6} \cdot I_S^{(R+1)}[v_2^{(R+1)}(e)]
\end{aligned}
$$

Processing the edge list is not required to reduce a source image $I_S^{(0)}$ of resolution $R = 0$ to a single target pixel I_T containing the average global background intensity.

$$
I_T \ := \ \tfrac{1}{12} \sum_{v=1}^{12} I_S^{(0)}[v]
$$

3.5.2 Image Expansion and Interpolation

Expanding a geodesic grid-based spherical image to the next higher level of resolution requires super-sampling at all vertex positions of the high resolution grid. Since a grid of resolution $R + 1$ contains all vertices of resolution R, the sampling positions of both grids coincide for all pixels of the low resolution image. These pixels can be copied directly into the target image, as indicated by a weighting factor of 1 in Fig. 3.10c.

All remaining vertices of the high resolution grid do not exist in the low resolution, therefore the corresponding pixels need to be determined by interpolation. Due to construction, each of these new vertices is located on a great circle arc connecting two vertices of the low resolution grid. Since the new vertex is equidistant from both old vertices, the two corresponding pixels of the low resolution image are weighted equally with a factor of $1/2$ (Fig. 3.10c). No distinction between pentagonal and hexagonal pixels is required for either copying or interpolating pixels.

Algorithm

A spherical source image $I_S^{(R)}$ of resolution level R is expanded to a target image $I_T^{(R+1)}$ of the next higher resolution level $R + 1$ in two processing steps. First, all pixels of the source image are copied into the first $|V^{(R)}|$ target pixels. Due to vertex sorting, pixels representing the same sampling position on the sphere have equal indices in both the source and target images.

for all vertices $v \in \{1, \dots, |V^{(R)}|\}$
$\quad I_T^{(R+1)}[v] \; := \; I_S^{(R)}[v]$

All remaining target pixels are located in the space between the sampling directions of the low resolution source image and need to be determined by interpolation. They are initialized using

for all vertices $v \in \{|V^{(R)}| + 1, \dots, |V^{(R+1)}|\}$
$\quad I_T^{(R+1)}[v] \; := \; 0$

and are accessed through the edge list of the high resolution grid. Due to edge sorting, only the first $\frac{1}{2}|E^{(R+1)}|$ edges are connected to a vertex of the low resolution source grid. Each edge is used only in one direction since only one of its vertices represents a source pixel. As pointed out above, each target pixel is interpolated from two equidistant source pixels, and thus each source pixel is weighted with a factor of $1/2$.

for all edges $e \in \{1, \dots, \frac{1}{2}|E^{(R+1)}|\}$
$\quad I_T^{(R+1)}[v_2^{(R+1)}(e)] \; := \; I_T^{(R+1)}[v_2^{(R+1)}(e)] + \frac{1}{2}I_T^{(R+1)}[v_1^{(R+1)}(e)]$

Interpolation is not required to expand a single source pixel I_S containing the average global background intensity to a spherical target image $I_T^{(0)}$ of resolution $R = 0$. The intensity value can be assigned directly to all 12 pixels.

for all vertices $v \in \{1, \dots, 12\}$
$\quad I_T^{(0)}[v] \; := \; I_S$

3.6 Local Motion Detection on Spherical Images

Visual motion detection provides a rich source of information about the self-motion of an observer as well as the three-dimensional structure of the environment (Gibson, 1950). Since motion detection is based on analyzing the changes occurring in a series of time-varying input images, it requires appropriate temporal filtering in addition to spatial filtering. This section describes efficient implementations of both temporal filtering and insect-inspired local motion detection on geodesic grid-based spherical images.

3.6.1   Temporal Filtering

The finite signal transduction speed and processing capacity of both natural and artificial vision systems limits the detection of temporal changes in a continuous stream
of images to a certain, maximum frequency. Faster changes cannot be resolved and
may lead to artifacts. In time-discrete systems, changes faster than $1/2$ of the sampling frequency result in temporal aliasing (Mallot, 2000; Jähne, 2001). In addition,
the incoming signal may be corrupted by high frequency photoreceptor and transduction noise. Both temporal aliasing and high frequency noise need to be suppressed by
a temporal low pass filter. On the other extreme, very low frequencies such as the average background signal (i.e., the ambient light level) and slow changes thereof do not
contribute useful information to visual motion detection and need to be removed by
temporal high pass filtering. In analogy to filtering in the spatial domain (Section 3.4),
a temporal low pass filter smooths the incoming signal by local averaging, i.e., by reducing the differences between successive values. A temporal high pass removes the
local average or DC component from the signal and thereby enhances the temporal
contrast, i.e., the changes between successive values.

   In the following, the function $I(t)$ denotes the time-varying intensity signal of a
single pixel or light receptor. A temporal filter is applied to this signal by convolution
with a suitable kernel defined by the function $K(t)$, and yields the filtered response
signal

$$F(t) = I(t) * K(t). \tag{3.16}$$

In contrast to spatial convolution, the kernel of a temporal filter cannot be isotropic,
i.e., symmetrically distributed around a maximum located at the current time t, since
the input signal $I(t')$ is not known for $t' > t$.[1] Therefore, Gaussian or DOG kernels
are not suitable for temporal low pass or high pass filtering.

   In this study the impulse response of a first-order temporal low pass filter is defined
by an exponential decay curve

$$K(t) = \left\{ \begin{array}{cc} 1/\tau \cdot (1 - 1/\tau)^t, & t \geq 0, \tau > 1 \\ 0, & t < 0 \end{array} \right. \tag{3.17}$$

with time constant τ. Using this function as a filter kernel yields a standard IIR (infinite impulse response; cf. Oppenheim and Willsky (1983) or Jähne (2001) for an
introduction) temporal low pass filter

$$L(I) = \left\{ \begin{array}{cc} \frac{1}{\tau} \int_{-\infty}^{t} (1 - \frac{1}{\tau})^{t-t'} \cdot I(t') \, dt', & \tau > 1 \\ I(t), & \tau = 1 \end{array} \right. \tag{3.18}$$

---

[1]However, an isotropic convolution kernel can be applied to a pre-recorded time series, where the
complete time course of the signal is known.

At each time t, this filter computes the local weighted average by integrating over the preceding signals $I(t')$ for $t' \leq t$. The sensitivity of the temporal receptive field is maximal for the most recent input signal $I(t)$, whereas preceding signals receive less weight.

For time-discrete applications, e.g., in digital image processing, the filter has the form

$$L(I) = \begin{cases} \frac{1}{\tau} \sum_{t'=-\infty}^{t} (1 - \frac{1}{\tau})^{t-t'} \cdot I_{t'}, & \tau > 1 \\ I_t, & \tau = 1 \end{cases} \quad (3.19)$$

where I_t denotes the intensity value of a single pixel at time step t. This filter is equivalent to the recursive definition

$$L(I_t) = (1 - \frac{1}{\tau}) \cdot L(I_{t-1}) + \frac{1}{\tau} \cdot I_t \quad (3.20)$$

$$= L(I_{t-1}) + \frac{1}{\tau} \cdot (I_t - L(I_{t-1})), \quad \tau \geq 1. \quad (3.21)$$

which dramatically reduces the amount of memory and computation time. At each time step, a single value, the filter response of the immediately preceding time step, and three basic arithmetic operations are sufficient to update the filter response. $1 - \frac{1}{\tau}$ and $\frac{1}{\tau}$ are constants and need to be computed only once.

In analogy to spatial high pass filtering described in Section 3.4, a simple temporal high pass filter $H(I_t)$ is obtained by subtracting the local average, i.e., the output of a temporal low pass filter $L(I)$, from the current input signal I_t.

$$H(I_t) = I_t - L(I_t) \quad (3.22)$$

Subtracting the output signals of two low pass filters L_1 and L_2 with different time constants τ_1 and τ_2 results in a temporal band pass filter.

$$B(I_t) = L_1(I_t) - L_2(I_t), \quad \tau_2 > \tau_1 \quad (3.23)$$

Algorithm

Temporal filtering of a geodesic grid-based source image $I_S^{(R)}$ of resolution R yields a target image $I_T^{(R)}$ of equal resolution. This is a strictly local operation computed separately for each pixel of the spherical image. The graph structure of the geodesic grid is not used since no information from adjacent pixels is required. Thus, the implementation is straightforward and requires only a single pass through the vertex list

```
for all vertices v ∈ {1,...,|V^(R)|}
    I_T^(R)[v]  :=  F(I_S^(R)[v])
```

with $F(I)$ representing an arbitrary temporal filter, e.g., a temporal low pass $L(I)$ or high pass $H(I)$.

3.6.2   Local Motion Detection

Self-motion of an observer through the environment induces characteristic patterns of image motion, the so-called optic flow (Gibson, 1950; Koenderink, 1986). The goal of local motion detection is to recover the direction and magnitude of the optic flow field at each position in the image from the changing intensity values. Although it is generally impossible to measure the true optic flow, i.e., the projection of the three-dimensional motion vector field in the environment onto the imaging surface[2] (Mallot, 2000), many useful algorithms have been suggested to estimate local image motion under certain constraints. The different algorithms are extensively discussed in textbooks on digital image processing for technical applications (Jähne, 2001) and on computational vision in a biological context (Mallot, 2000).

Implementing the Reichardt Elementary Motion Detector

This study uses a delay-and-correlate motion detection scheme derived from behavioral experiments with insects (Hassenstein and Reichardt, 1956; Reichardt, 1969; Egelhaaf and Borst, 1993). The computational properties of the so-called Reichardt elementary motion detector (EMD) are well understood (Poggio and Reichardt, 1973; Borst and Egelhaaf, 1989, 1993). As illustrated in Fig. 3.11, a Reichardt-type EMD is composed of two mirror-symmetrical semidetectors sharing two input signals I_1 and I_2 from separate locations in the input image. Each semidetector correlates a delayed version $L(I)$ of one of the input signals with the other, undelayed signal. The temporal delay stage is often modeled as a first-order IIR temporal low pass filter (Eq. 3.18), exploiting the phase shift in the signal introduced by this type of filter (Fig. 3.12). The subsequent correlation stage requires a nonlinear interaction, e.g., a multiplication, between the delayed and undelayed signals from both sides (Borst and Egelhaaf, 1989). The semidetector responses are subtracted to yield the full detector output signal

$$\Phi(I_1, I_2) = L(I_1) \cdot I_2 - I_1 \cdot L(I_2). \tag{3.24}$$

Each EMD detects a one-dimensional component of local image motion along its preferred direction which is defined by the orientation of the line connecting both sampling points of the EMD in the image. The highest response signal is obtained when the direction of image motion is aligned with the orientation of the detector. Image motion orthogonal to the receptor orientation yields no response. The sign of the output signal is reversed when the image moves against the preferred direction. If both subunits of an EMD are balanced, i.e., both semidetectors have delay stages with equal time constants and are subtracted with equal weight, the absolute values of the response signals for motion in and against the preferred direction are equal.

---

[2]For instance, it is not possible to visually determine the rotation of an untextured sphere with a Lambertian surface.

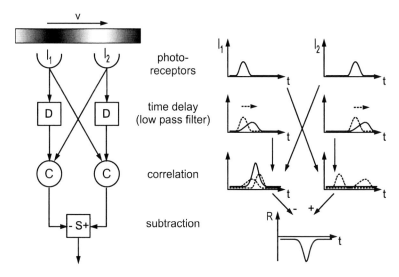

Figure 3.11: Simplified correlation-type elementary motion detector after Hassenstein and Re-ichardt (1956). Two neighboring photoreceptors detect the light intensities I_1 and I_2. In each of two mirror-symmetric semidetectors one of the input signals is delayed by a temporal low pass filter and subsequently correlated with the other, undelayed signal. The detector response is maximal when the image displacement during the temporal delay equals the angular distance between the two photoreceptors. The sign of the difference between both semidetector responses indicates the direction of motion.

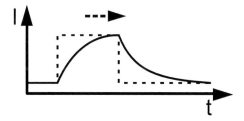

Figure 3.12: Phase shift from temporal low pass filtering. A standard IIR temporal low pass filter introduces a phase lag in the signal and can therefore be used as a temporal delay unit with additional signal smoothing. The input signal is shown as dashed line, and the output as solid line.

The response of a Reichardt-type EMD is ambiguous and does not represent true image velocity. It is maximal when the image displacement during the temporal delay equals the angular distance between the two sampling points, and smaller for both slower and faster image motion. Thus, the parameters determining the image velocity where the detector response is maximal include the base angle, i.e., the spatial distance between the two sampling points, and the time constant of the temporal delay stage. Despite the ambiguity with respect to true image velocity, the sign of the average EMD response reliably indicates the direction of motion, and for small image velocities the average absolute value of the detector response increases almost linearly.

Further, the output signal of a single EMD depends on the phase, contrast, and spatial frequency content of the input pattern. In particular, low spatial frequencies cause transients and oscillations in the output signal. However, the dependence on the input pattern can be reduced by spatial integration over an array of multiple EMDs looking at different phases of the stimulus pattern (Borst and Egelhaaf, 1989).

Spherical Arrays of EMDs on Geodesic Grids

Geodesic grids provide an elegant and efficient way of homogeneously arranging large numbers of Reichardt-type EMDs on the sphere. Assigning the EMDs to the edges of a geodesic grid ensures that each detector receives its input signals from two adjacent pixels of the spherical image. The base angle between the two input viewing directions of each EMD therefore equals the arc length of the corresponding edge, and its preferred direction coincides with the local orientation of the edge. Thus, each edge samples the spherical optic flow field at a specific location and local orientation. The resulting flow field estimate contains one detector response signal per edge, each representing a one-dimensional component of local image motion. Since most pixels have six neighbors, the optic flow can be locally estimated in six different directions in a highly redundant manner. A distinction between pentagonal and hexagonal pixels is not required since an edge always connects two vertices regardless of the number of their neighbors.

Geodesic grid-based arrays of EMDs offer a number of useful properties for spherical motion detection:

1. The locations and the preferred directions of the individual EMDs are homogeneously distributed on the sphere.

2. All EMDs have a quasi-uniform base angle since the arc lengths of the edges in a geodesic grid are approximately equal (cf. Tab. 3.2).

3. The preferred directions of the six EMDs connected to a particular pixel are separated by approximately equal angles, i.e., they are distributed around the pixel in a locally isotropic manner.

4. The grid of EMDs covers the entire sphere in a seamless manner without special cases or polar singularities.

5. The information in the spherical input image is efficiently exploited since each pixel is used by five or six different EMDs.

Algorithm

Local motion detection based on insect-inspired correlation-type EMDs can be efficiently implemented on spherical geodesic grids as a sequence of two processing stages. At each time step, the first stage updates a temporally delayed version $I_L^{(R)}$ of the current spherical source image $I_S^{(R)}$ by applying a standard IIR temporal low pass filter (Eq. 3.21) to each pixel. This requires one pass through all vertices of the geodesic grid.

$$\text{for all vertices } v \in \{1, \ldots, |V^{(R)}|\}$$
$$I_L^{(R)}[v] := I_L^{(R)}[v] + \frac{1}{\tau}(I_S^{(R)}[v] - I_L^{(R)}[v])$$

The second processing stage comprises two local correlations and one subtraction for each EMD. Each semidetector correlates the temporally delayed signal from one of its inputs with the undelayed signal from the other input. The input pixels are represented by the two vertices $v_1^{(R)}(e)$ and $v_2^{(R)}(e)$ of the edge e on which the EMD is located. The asymmetric time delay between the input signals is implemented by taking one of the signals from the delayed image $I_L^{(R)}$ whereas the other signal is taken from the undelayed image $I_S^{(R)}$. Thus, each pixel of both images serves as input for five or six different EMDs, depending on the number of its neighbors. Finally, subtracting both semidetector responses yields the spherical target image $\Phi_T^{(R)}$ containing the EMD output signals. All three local operations are applied to all EMDs in a single pass through the edge list.

$$\text{for all edges } e \in E^{(R)}$$
$$\Phi_T^{(R)}[e] := I_L^{(R)}[v_1^{(R)}(e)] \cdot I_S^{(R)}[v_2^{(R)}(e)] -$$
$$I_S^{(R)}[v_1^{(R)}(e)] \cdot I_L^{(R)}[v_2^{(R)}(e)]$$

Note that the resulting spherical target image $\Phi_T^{(R)}$ represents the EMD output signals on the edges of the geodesic grid, whereas the pixels of the spherical input images $I_S^{(R)}$ and $I_L^{(R)}$ are stored on the vertices.

3.7   Results

This section shows various examples of geodesic grid-based spherical image processing using the algorithms described above, as well as the update rates and memory requirements on a standard PC. The spherical input images were acquired in a realistic 3D virtual environment using the omnidirectional compound eye simulation model described in Chapter 2.

3.7.1   Examples

Generic Convolution on the Sphere

As described in Section 3.3.2, geodesic grid-based spherical images can be convolved with filter kernels of arbitrary sizes. The filtering is homogeneous on the entire sphere and therefore free of polar singularities or other distortions introduced by inappropriate image representations. The examples shown in Fig. 3.13*b*, *c*, *e* and *g* result from applying different Gaussian kernels with standard deviations between $1.25°$ and $10°$ and filter mask diameters between $5°$ and $40°$ to a spherical image of resolution $R = 5$. The input image is shown in Fig. 3.13*a*. As in the planar case, local averaging eliminates high spatial frequencies, and the resulting low pass filtering or smoothing of the image increases with the width of the filter kernel.

Examples of spherical convolutions of the same input image with different DOG kernels are shown in Fig. 3.13*d*, *f* and *h*. The DOG kernels have positive Gaussian components of $2.5°$, $5°$ and $10°$ standard deviation, and negative components of $1.25°$, $2.5°$ and $5°$. The filter mask diameters are $10°$, $20°$ and $40°$, respectively. The resulting images demonstrate the band pass characteristics of DOG filters: The local background signal is strongly reduced whereas contours and edges are enhanced. Note that convolution is a linear operation, therefore it is equivalent to implement a band pass filter by either convolving the image directly with the DOG kernel (which was used here) or by subtracting the images resulting from two separate convolutions with both Gaussian components of the DOG kernel.

Graph-Based Convolution

The graph-based convolution algorithm described in Section 3.4 applies filter masks with a diameter of three pixels homogeneously to the entire sphere. This allows to implement biologically inspired local interactions of pixels with their immediate neighbors such as local averaging and lateral inhibition. Fig. 3.14 demonstrates spatial low pass and high pass filtering using the isotropic convolution kernels shown in Fig. 3.9. The graph-based convolution is applied separately to each of six spherical input images with different resolution levels $R \in \{0, \ldots, 5\}$ depicted in the left column of Fig. 3.14.

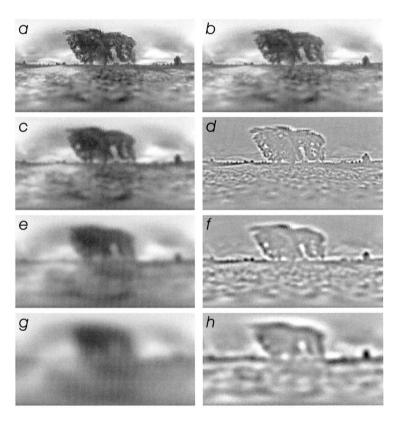

Figure 3.13: Generic convolution of spherical images. *a.* Original image ($R = 5$). *b,c,e,g.* Gaussian low pass filtering with standard deviations $1.25°$, $2.5°$, $5°$ and $10°$, and filter mask diameters $5°$, $10°$, $20°$ and $40°$, respectively. *d,f,h.* DOG band pass filtering with positive components $2.5°$, $5°$ and $10°$, negative components $1.25°$, $2.5°$ and $5°$, and filter mask diameters $10°$, $20°$ and $40°$, respectively.

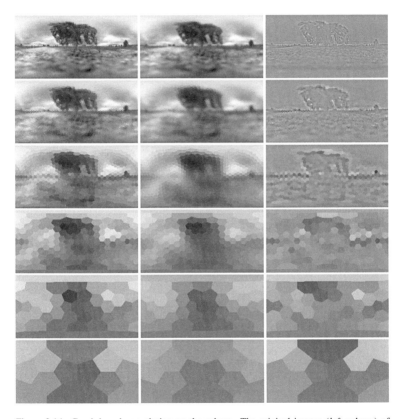

Figure 3.14: Graph-based convolution on the sphere. The original images (*left column*) of different resolution levels $R \in \{0, \ldots, 5\}$ are low pass (*center column*) and high pass (*right column*) filtered using filter masks with a diameter of three pixels (Fig. 3.9).

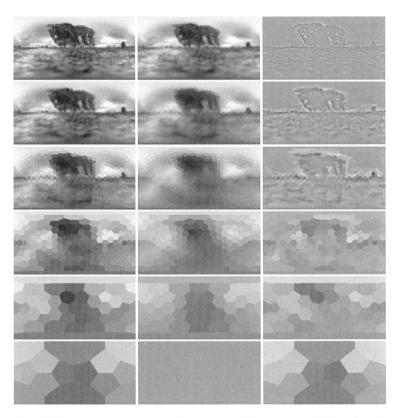

Figure 3.15: Spherical resolution pyramids. A spherical Gaussian pyramid (*left column*) is computed by repeated reduction of the original image ($R = 5$, *upper left corner*). Each image in the Gaussian pyramid is expanded and interpolated to the next higher level of resolution (*center column*). Subtracting each pair of images with equal resolution yields a spherical Laplacian pyramid (*right column*).

The resulting low pass and high pass filtered images are shown in the center and in the right column, respectively.

Spherical Resolution Pyramids

Algorithms for the efficient computation of geodesic grid-based spherical resolution pyramids are introduced in Section 3.5. Here, these algorithms are applied to an exemplary spherical image of resolution $R = 5$ shown in the upper left corner of Fig. 3.15. Starting with this input image, a spherical Gaussian pyramid is computed by repeatedly reducing the image to a geodesic grid of lower resolution (*left column*). The low pass filter mask depicted in Fig. 3.9 is applied to each pixel to prevent spatial aliasing due to sub-sampling. The lowest resolving image of the Gaussian pyramid is shown in the bottom image of the left column of Fig. 3.15. It contains 12 pixels corresponding to the 12 vertices of an undivided icosahedron ($R = 0$). The final reduction step yields a single pixel containing the global average color of the image (not shown in Fig. 3.15).

Expanding each image of the Gaussian pyramid to the next higher level of resolution yields a smoothed image pyramid depicted in the center column of Fig. 3.15. The uniform color of the image at the bottom of the center column results from expanding the global average color to an image of resolution $R = 0$.

The right column of Fig. 3.15 shows a spherical Laplacian pyramid. It contains a set of band pass filtered images of different resolutions which are obtained by subtracting each image of the Gaussian pyramid (*left column*) from the corresponding image in the expanded Gaussian pyramid (*center column*).

A comparison of Fig. 3.15 with Fig. 3.14 shows that the expanded Gaussian pyramid and the Laplacian pyramid closely approximate a true spherical convolution with a Gaussian and a DOG kernel, respectively. In analogy to planar image pyramids (Jähne, 2001), slight differences between corresponding images of both figures are due to the fact that computing a Gaussian pyramid discards information by sub-sampling the images to reduce their resolution. Since this information cannot be reconstructed when the images are subsequently expanded to their original resolution, it needs to be substituted by interpolated values.

Local Motion Detection

Homogeneous local motion detection on spherical images is demonstrated in Fig. 3.16 for both clockwise roll rotation (*left column*) and forward translation (*right column*) of a virtual observer. For each example a sequence of 250 spherical input images of resolution level $R = 3$ is recorded using the omnidirectional compound eye simulation model described in Chapter 2. The final images of each sequence are depicted in Fig. 3.16*a* and *b*.

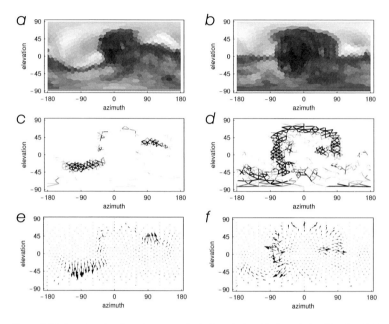

Figure 3.16: Local motion detection on a spherical geodesic grid of resolution $R = 3$. *a,b*. Input image acquired during roll rotation (*a*) and forward translation (*b*) of the observer. *c,d*. Output signals of 1920 correlation-type local EMDs located on the edges of the grid. Darker lines indicate higher absolute values. *e,f*. Corresponding optic flow fields computed from the EMD output signals.

The 642 individual pixels of the image sequences serve as input signals for 1920 insect-inspired correlation-type local EMDs distributed over the sphere. As pointed out in Section 3.6.2, each EMD resides on an edge of the geodesic grid and thereby connects two adjacent pixels of the spherical image. The edges of the geodesic grid sample the image motion in a highly uniform manner over the entire sphere since each EMD has an approximately equal base angle of $8.64°$ (cf. Tab. 3.2) and the local orientations of the individual EMDs are homogeneously distributed over the sphere. Fig. 3.16*c* and *d* show the output signals of the individual EMDs for the final images of the example sequences. The orientation of each line indicates the local EMD direction. The gray level indicates the scaled absolute value of the output signal, with darker lines representing higher values. Note that each signal may have a positive or negative

Filter Mask	Update Rate [s^{-1}]			
Total Width	$R = 4$		$R = 5$	
[°]	Gray	Color	Gray	Color
5	1066.7	336.8	160.0	61.5
10	711.1	220.7	114.3	39.8
20	453.9	141.9	42.1	14.0
40	152.0	51.1	11.4	3.8

Table 3.4: Performance of spherical convolution. Filter masks of different diameters are applied to discrete spherical gray value and RGB color images of resolution levels $R = 4$ and $R = 5$. The update rates are determined on a standard PC with Intel Pentium III 850 MHz CPU.

sign, indicating forward or backward image motion along the preferred direction of the corresponding EMD.

Geodesic grid-based spherical motion fields can be converted to a conventional, two-dimensional vector field representation by integrating all local motion components connected to each vertex. The vector fields computed from the EMD output signals are depicted in Fig. 3.16e and f. Each arrow indicates the local direction and magnitude of the flow field at one position on the sphere.

Both examples demonstrate that correlation-type EMDs are most sensitive to moving edges with high contrast. In the rotation sequence the maximum local response is induced by the rotating horizon, whereas in the translation sequence the largest output signals are generated for the edges of the approaching obstacle. This response characteristic is consistent with both biological findings and theoretical considerations (Götz, 1964; Reichardt, 1969; Egelhaaf and Borst, 1993; Borst and Egelhaaf, 1993).

3.7.2 Performance

All algorithms for geodesic grid-based spherical image processing were implemented and tested on a standard PC with an 850 MHz Intel Pentium III CPU. The input and output images use floating point numbers to represent gray values, RGB color components, and local motion signals.

Generic Spherical Convolution

The performance of generic spherical convolution is shown in Tab. 3.4. Using the algorithm described in Section 3.3.2, four different circular filter masks with diameters $5°$, $10°$, $20°$ and $40°$ are applied to spherical images of resolution $R = 4$ and $R = 5$. The actual convolution kernels may have arbitrary sensitivity distributions within the

R	Update Rate [s^{-1}]		
	Spatial	Temporal	Motion
0	457143.4	2628010.1	718734.2
1	45649.1	956730.8	192481.7
2	11173.2	267363.9	48808.4
3	2764.3	68175.8	12260.5
4	687.4	17250.7	3043.3
5	153.0	4210.5	371.6

Table 3.5: Performance of graph-based local image processing on the sphere. Different local operations are applied to discrete spherical images of resolution levels $R \in \{0, \ldots, 5\}$. For spatial and temporal filtering (*left and center column*) one local operation is applied to each vertex, using a convolution kernel of three pixels diameter and a standard IIR (infinite impulse response) temporal low pass filter, respectively. For local motion detection (*right column*) the output signal of one correlation-type EMD, including a temporal low pass filter for time delay, is computed for each edge. Testing environment as in Tab. 3.4.

filter mask diameter. Examples of spherical images of resolution $R = 5$ filtered with different Gaussian and DOG kernels are depicted in Fig. 3.13. The update rates are shown for both gray value and RGB color images. As expected, gray value images are processed approximately three times faster than color images since they contain only $1/3$ of the amount of data.

For most filter masks the update rates are sufficient for real-time spherical image processing at standard video frame rates of 50 or 60 Hz. However, applying large filter masks to high resolution images can be considerably accelerated by reducing the image resolution prior to filtering, e.g., by using a spherical resolution pyramid.

Graph-Based Spherical Image Processing

Tab. 3.5 shows the update rates of different local image processing operations based on the graph structure of geodesic grids. Spherical input images of six resolution levels $R \in \{0, \ldots, 5\}$ are used to demonstrate graph-based convolution, temporal filtering and local motion detection. The algorithms are described in Sections 3.4, 3.6.1 and 3.6.2, respectively.

Graph-based spherical convolution (*left column*) applies hexagonal and pentagonal filter masks with a diameter of three pixels to each vertex of the geodesic grid. Example output images of resolution $R = 5$ are depicted in Fig. 3.14. Temporal filtering (*center column*) updates a standard IIR (infinite impulse response) low pass filter (Oppenheim and Willsky, 1983) for each vertex. The update rates for temporal high pass filtering are almost identical (data not shown). Local motion detection (*right column*) is computed

| Pyramid Type | Update Rate $[\mathrm{s}^{-1}]$ | |
($R = 0..5$)	Gray	Color
Gaussian	496.8	150.2
Laplacian	197.8	59.4

Table 3.6: Performance of spherical resolution pyramids. All six resolution levels ($R = 0..5$) of a Gaussian and a Laplacian pyramid are computed for both gray value and RGB color images, starting with an image of resolution $R = 5$. The update rate for the Laplacian pyramid includes the computation of the corresponding Gaussian pyramid. Testing environment as in Tab. 3.4.

for each edge of the geodesic grid using insect-inspired correlation-type elementary motion detectors. The EMD update rates include a run of a temporal low pass filter for time delay on the vertices of the input image. Fig. 3.16 shows examples of input and output images for geodesic grid-based local motion detection on the sphere at a resolution level of $R = 3$.

All graph-based local filtering operations achieve very high update rates even at high resolution levels, thus they are fully suitable for real-time spherical image processing.

Spherical Resolution Pyramids

The update rates of geodesic grid-based spherical resolution pyramids introduced in Section 3.5 are presented in Tab. 3.6. All six resolution levels $R = 0..5$ of both a Gaussian and a Laplacian pyramid are computed from a single spherical input image of resolution $R = 5$. Fig. 3.15 shows the resulting images for both pyramids.

At 496.8 updates per second, the computation of a complete spherical Gaussian pyramid is more than 3 times faster than a convolution of the original image alone (153.0 updates per second, cf. Tab. 3.5). Identical low pass filter masks (Fig. 3.9) are used in both cases. The complete Laplacian pyramid including the computation of the Gaussian pyramid and the subsequent expansion and subtraction of all images achieves 197.8 updates per second and is still 1.3 times faster than the convolution of only the highest level of resolution. Thus, whenever an approximation of a true convolution with a Gaussian or DOG kernel (cf. Section 3.5) is sufficient for a given application, spherical low pass and high pass filtering can be considerably accelerated by replacing the convolution by image reduction and expansion.

The update rates of the six-level spherical resolution pyramids shown here are fully sufficient for real-time processing at typical camera frame rates used in computer vision or robotic applications.

R	Memory Requirements [kB]			
	Vertex Image		Edge Image	
	Single	Pyramid	Single	Pyramid
0	0.05	0.05	0.12	0.12
1	0.16	0.21	0.47	0.59
2	0.63	0.84	1.88	2.46
3	2.51	3.35	7.50	9.96
4	10.01	13.36	30.00	39.96
5	40.01	53.37	120.00	159.96
6	160.01	213.38	480.00	639.96

Table 3.7: Memory requirements for geodesic grid-based spherical images.

3.7.3 Memory Requirements

The amount of memory required to store spherical images on geodesic grids of different resolution levels is shown in Tab. 3.7. Image information can be stored on the vertices (*left*) and on the edges (*right*) of a geodesic grid. The memory calculation assumes that four bytes are used for each pixel, e.g., one floating point number representing a local intensity or a local motion signal. Storing an entire spherical image pyramid including all resolution levels requires only an additional $1/3$ of the memory used by the highest level of resolution alone.

3.8 Discussion

In this chapter I introduced geodesic grids as a novel data structure for the representation and processing of discrete spherical images. A spherical image is defined as a set of discrete data points located at the vertices of a recursively subdivided icosahedron, resulting in a raster of hexagonal pixels. Directional information such as local optic flow is represented on the edges of the grid. In contrast to traditional, planar representations such as catadioptric or Mercator projections, geodesic grids are homogeneous in the distribution of sampling directions over the sphere and isotropic in the connectivity and distance of adjacent pixels. Moreover, they do not contain poles or other singularities, and no resolution is wasted in any region of the sphere. An image is stored in computer memory as a single, one-dimensional vector of gray or color values. Individual pixels are efficiently accessed using the underlying graph structure, allowing various biologically plausible image processing operations such as local averaging, lateral inhibition, local motion detection by correlation-type EMDs, as well as the construction of spherical resolution pyramids. All of these operations are applied homogeneously and in a seamless manner to the entire spherical image.

Below I will discuss applications of geodesic grid-based spherical image processing in insect biology and machine vision.

3.8.1   Modeling Insect Vision with Geodesic Grids

The arrangement of sampling directions as well as the local image processing operations described in this chapter closely resemble the type of image representation and processing found in the visual system of insects. Thus, in biological studies on insect vision and visually controlled behavior geodesic grids can serve as an approximate model for the spherical distribution of receptor units in insect compound eyes. In addition, they can provide the underlying data structure for modeling the subsequent, retinotopic processing layers including retina, lamina, medulla, lobula and lobula plate.

The most striking similarities between insect compound eyes and icosahedral geodesic grids are the spherical field of view and the arrangement of discrete sampling points on a hexagonal lattice. The similarities and differences between both types of sampling grids will be discussed in the following.

Spherical Field of View

In insects the spherical field of view is divided into a left and right hemisphere by separate compound eyes and optic lobes for each side (cf. Section 3.1.1), whereas geodesic grids cover the entire sphere. In most applications this apparent difference can be neglected since (1) the diameter of insect eyes is very small compared to typical object distances during flight, allowing to assume a singular viewpoint of all local receptor viewing directions (Schwind, 1989), (2) all image processing operations in the retinotopic layers are local and therefore independent of the particular hemisphere, and (3) full spherical images on geodesic grids allow the definition of wide-field integration units with arbitrary receptive fields, including inter-hemispherical interactions.

In some species such as *Drosophila* the field of view does not cover the entire sphere but has a blind zone occluded by the body, whereas the visual fields of both eyes may overlap in the frontal region (cf. Fig. 2.2). Since stereo vision during flight can be excluded due to the small eye diameter (Schwind, 1989), binocular overlaps can be simulated using a single spherical input image. Blind zones can be modeled by simply ignoring all input pixels in the corresponding region.

Distribution of Sampling Points and Local EMDs

The vertices of geodesic grids form a quasi-uniform, hexagonal lattice of sampling points on the sphere which closely resembles the layout of compound eyes. However, the number of vertices in a geodesic grid is determined by its resolution level. It is not possible to generate geodesic grids for arbitrary numbers of pixels, e.g., to match the exact number of ommatidia in a specific insect eye. The resolution level with the closest fitting number of pixels needs to be used instead. The range of resolutions occurring in insect compound eyes is covered by the four levels $R \in \{3, \ldots, 6\}$. Further, each geodesic grid contains 12 pentagonal pixels and exhibits the same axes of symmetry as the original icosahedron. Insect compound eyes also contain receptor units with less than six neighbors (Petrowitz et al., 2000), but the global arrangement of rows in the hexagonal lattice differs from the layout of geodesic grids. It is not yet known whether the orientation of the hexagonal grid has a specific function or is optimized for a specific behavior (Petrowitz et al., 2000). Many compound eyes exhibit acute zones of enhanced resolution (cf. Section 2.1.1) which are not present in geodesic grids. As demonstrated in Fig. 2.14, acute zones can be simulated by appropriately distorting the vertex positions of a geodesic grid. Since moving the vertex positions does not change the topology of the sampling grid, this provides an elegant method for implementing space-variant, spherical image processing.

Placing local motion detectors on the edges of an icosahedral geodesic grid closely resembles the EMD arrangement in the insect visual system. In the light adapted eye, motion detection is dominated by interactions between directly adjacent photoreceptors (Kirschfeld, 1972; Buchner, 1976; Riehle and Franceschini, 1984) which are arranged on a hexagonal lattice covering an approximately spherical field of view (Buchner, 1976; Petrowitz et al., 2000). Both geodesic grids and insect eyes sample the image motion at each location on the sphere in a highly redundant manner along six different local orientations.

In conclusion, despite minor differences geodesic grids closely approximate the spherical sampling and processing lattice of the insect visual system, whereas other approaches based on Cartesian or polar grids (see Section 3.1.2) introduce space-variant distortions and singularities.

Image Processing Operations

The neuronal structures and processes contributing to image processing in the insect visual system are far more complex than the exemplary algorithms described in this chapter. They involve local adaptation and feedback as well as saturation and other nonlinear operations. Many cell types in the medulla, lobula and lobula plate have not yet been investigated in electrophysiological studies, and many functions are not yet completely understood. For instance, the exact location and neuronal implementation of the elementary motion detectors has not yet been identified, and there is evidence that there may be more than one system for visual motion detection (Srinivasan, 1993; Douglass and Strausfeld, 2001; Zanker, Srinivasan, and Egelhaaf, 1999). Geodesic grids provide an excellent basis for implementing and testing arbitrary models of local image processing in the insect visual system homogeneously on the sphere.

3.8.2   Geodesic Grids in Spherical Machine Vision

Omnidirectional imaging systems have a number of advantages compared to standard cameras with a limited field of view. All objects in the surrounding environment are simultaneously visible at all times, thus exploratory camera movements are not required to detect objects such as landmarks, obstacles, interaction partners, or potential hazards. Multiple objects can be tracked simultaneously, whereas a standard camera following one particular object may lose other objects as soon as they are no longer in view. Omnidirectional vision is especially useful for the detection and estimation of self-motion. It allows to reliably separate rotatory from translatory optic flow fields (Nelson and Aloimonos, 1988; Gluckman and Nayar, 1998). Moreover, the focus of expansion and the focus of contraction are always in view.

Geodesic grids allow full spherical image acquisition and processing in a highly efficient manner. The high update rates for spatial and temporal filtering as well as for local motion detection even on consumer-level computers (cf. Section 3.7.2) are fully sufficient for real-time applications in surveillance and robotics. For image resolutions comparable to insect compound eyes, multiple successive processing stages can be updated at standard video frame rates.

Image Acquisition

As demonstrated earlier in this chapter, homogeneous spherical image processing on geodesic grids has many advantages compared to traditional approaches. But how can geodesic grid-based images be acquired from the environment? In computer simulations, the most straightforward method is ray tracing, i.e., finding the closest intersection point of each local viewing direction with the objects in the environment and determining the local object color. Although this allows true spherical sampling along arbitrary directions, it has two severe disadvantages: (1) Point sampling leads to spatial aliasing, therefore multiple rays are required to average over a certain area around each local viewing direction. (2) At present, ray tracing is not directly supported by standard graphics hardware. To avoid these problems, the compound eye simulation model described in Chapter 2 generates multiple planar, perspective views covering the surrounding environment, and transforms them into discrete spherical images in a separate processing step. This method is very efficient since rendering perspective views is hardware-accelerated, and the subsequent filtering is based on a pre-computed lookup table. The compound eye simulation model was used to generate the spherical example images shown in this chapter.

A similar two-stage approach can be used to acquire geodesic grid-based spherical images in real-world applications such as robot navigation. First, an arbitrary omnidirectional imaging system (e.g., a wide-angle camera, a catadioptric system, or a multi-camera cluster) generates one or multiple planar images containing the entire spherical view in a distorted form. The raw images are subsequently transformed into a geodesic grid-based spherical representation using appropriately distorted, overlapping Gaussian filter masks. In analogy to the computer simulation method (Neumann, 2002), the entire transformation can be pre-computed and stored in a lookup table. All further image processing can be performed on the homogeneous, geodesic grid-based representation. In contrast, space-variant image processing is required each time a filtering operation is applied directly to a raw catadioptric image (Daniilidis et al., 2002) or to a Mercator projection computed from a catadioptric input image in an additional processing step (Chahl and Srinivasan, 2000).

A promising approach for a novel, omnidirectional imaging device inspired by the insect visual system is to arrange individual glass fibers along the sampling directions of a geodesic grid (or any other distribution of viewing directions), and bundle the opposite endings of the fibers on a CCD or CMOS camera chip. This provides a direct method to sample discrete spherical images from the environment and allows to define an arbitrary viewing direction and origin for each individual receptor. Omnidirectional imaging with fiber optics may be particularly interesting for the development of small, integrated hardware implementations of spherical vision and motion detection, e.g., using analog VLSI implementations of insect-inspired elementary motion detectors (Harrison and Koch, 1999, 2000; Indiveri and Douglas, 2000).

Applications

A spherical field of view is useful in numerous applications including surveillance, tracking, immersive multimedia, as well as robot vision and navigation. In principle, geodesic grids can be used in all of these applications to replace conventional approaches based on planar representations of spherical images, resulting in lesser memory requirements and faster processing due to homogeneous sampling (cf. Section 3.1.2).

Geodesic grids are particularly well-suited for the design of biomimetic, spherical image sensors and processing algorithms since they closely approximate the sampling lattice of insect compound eyes. Further, they allow the efficient implementation of local image processing operations such as spatiotemporal low pass and high pass filtering by local averaging and lateral inhibition, as well as local motion detection using correlation-type EMDs. As explained in Section 3.1.1, this corresponds to the type of image processing performed by the early, retinotopic layers of the insect visual system. Flying insects convincingly demonstrate their superiority over present technology in a multitude of complex spatial behaviors such as three-dimensional self-motion control, range finding, collision avoidance and navigation. Thus, insect-inspired spherical image acquisition and processing may be crucial for the development of simple but highly efficient visual control strategies for autonomous flying systems, especially in applications with limited computing capacities due to restrictions in weight and energy consumption, such as aerospace and miniature robotics.

In the next chapter I will use geodesic grid-based spherical image acquisition and processing to implement insect-inspired, visual orientation strategies for a fully autonomous, three-dimensional flight control.

Chapter 4

Learning to Fly - Behavior-Oriented Vision for Biomimetic Flight Control

Flying insects extract information about their spatial orientation and self-motion from visual cues such as patterns of light intensity or optic flow in a spherical field of view. In this chapter I present an insect-inspired neuronal filter model and show how optimal receptive fields for the detection of flight-relevant input patterns can be derived directly from the visual stimulus occurring during typical flight behavior. Using a least squares principle, the receptive fields are optimally adapted to all behaviorally relevant, invariant properties of the agent and the environment. Open-loop and closed-loop simulations in a highly realistic virtual environment show that four independent, purely reactive mechanisms based on optimized receptive fields for attitude control, course stabilization, obstacle avoidance and altitude control are sufficient for a fully autonomous visual flight stabilization with all six degrees of freedom.

4.1   Introduction

Behavior-oriented vision has its origin in J. J. Gibson's ecological theory of perception (Gibson, 1950). It is based on the observation that the behavior of an organism elicits characteristic patterns in the perceived stimulus. Identifying and exploiting these patterns allows to extract only such information from the environment which is required to control a specific behavior, and thereby to minimize the required amount of computation and internal representation. In particular, certain reactive behaviors can be immediately controlled by an appropriately selected visual input. For instance, phototropic reactions can be observed even in the most primitive animals, and were the first to be implemented in biologically inspired autonomous robots (Walter, 1950). Braitenberg (1984) showed that the interaction of apparently simple, reactive control mechanisms with the environment may lead to complex behavior.

Since natural evolution adapts the entire organism to specific behaviors in specific environments, analyzing biological examples may provide useful ideas and inspiration for the design of artificial systems on various levels, including sensor morphology, processing stages, control mechanisms, and behavior. Insect vision and behavior have been investigated for over a century and many of the mechanisms involved are well understood. Since flying insects are highly optimized for flight behavior in their natural environments, many aspects of insect vision and information processing can be imitated to control an artificial flying system in a comparable environment, and do not need to be re-invented or developed *de novo*.

Various insect-inspired, reactive control mechanisms have been implemented in robots and computer simulations. Visual tracking was investigated in a simulation study by Cliff (1992). Mura and Franceschini (1994) simulated altitude control and vertical obstacle avoidance behavior, assuming fixed attitude angles and pure forward translation in a vertical plane. Weber, Venkatesh, and Srinivasan (1997) implemented a horizontal centering response and speed regulation on a mobile robot. One-dimensional tracking and orientation towards a strong contrast was demonstrated by Huber et al. (1999). However, all of these systems were limited to motion in a horizontal or vertical plane with one or two degrees of freedom. Neumann and Bülthoff (2001) showed that a basic three-dimensional flight stabilization can be achieved by a combination of four insect-inspired visual orientation strategies, including the dorsal light response for attitude control (Hengstenberg et al., 1986), the optomotor response for course stabilization (Reichardt, 1969; Götz, 1968; Hausen and Egelhaaf, 1989), as well as visual range finding from translatory image motion for obstacle avoidance (Tammero and Dickinson, 2002; Srinivasan, 1993) and altitude control (Srinivasan et al., 2000).

In this chapter I demonstrate how the behavior-oriented extraction of visual information from the environment by wide-field integration units can be optimized in the sense of least squares, exploiting the covariance of input signals and behavior. In contrast to an earlier study on visual self-motion estimation (Franz et al., 1999), the optimal receptive fields are derived directly from the visual input and do not require explicit modeling of the distance or contrast distributions in the environment. For all four orientation strategies the resulting optimal receptive fields as well as their open-loop tuning curves are presented and discussed. Furthermore, examples for closed-loop flight behavior are shown to demonstrate that these strategies are capable of stabilizing an airborne system with all six degrees of freedom in a highly realistic virtual environment.

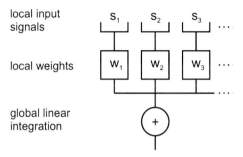

Figure 4.1: Linear wide-field integration unit. The unit receives input signals from a large population of local light receptors or elementary motion detectors. It is tuned to a specific input pattern determined by the distribution of local weights.

4.2 Flight-Relevant Information from Visual Cues

4.2.1 Modeling Wide-Field Integration Units

Insects use wide-field integration units to evaluate the local light intensity and local optic flow signals from preceding retinotopic processing steps. These units have complex receptive fields which are highly selective for behaviorally relevant input patterns. Their sensitivity distributions contain prior knowledge on invariant properties of both the behavior and the environment. Examples are the tangential neurons in the blowfly *Calliphora erythrocephala* which are tuned to characteristic, self-motion-induced optic flow fields (Krapp and Hengstenberg, 1996; Krapp et al., 1998; Franz and Krapp, 2000).

A model for wide-field integration units is depicted in Fig. 4.1. A large population of locally weighted input signals is spatially integrated to yield the receptive field response r as the scalar product

$$r = \sum_i w_i s_i = \mathbf{w}\mathbf{s} \qquad (4.1)$$

of the input signal vector $\mathbf{s}$ and the weight vector $\mathbf{w}$. To control the different components of flight behavior, various integration units evaluate the input signal distribution in parallel, each unit tuned to a specific pattern determined by the particular distribution of local weights.

Based on this general processing architecture inspired by the fly visual system, the exact weight distributions can be determined using an optimization method. In the following I show how optimal global receptive fields are derived directly from the local receptor signals occurring during typical behavior. Using a least squares principle, the sensitivity distributions are optimally adapted to all behaviorally relevant, invariant properties of the agent and the environment, such as systematic errors in the motion detector signals, typical movements of the agent, and characteristic distributions of light intensity, color, contrast and distance in the environment.

4.2.2   Optimal Receptive Fields

Let $\mathbf{r} \in \Re^m$ be the vector of activation signals for a single motor unit, recorded at m measurements during typical behavior, and $\mathbf{S} \in \Re^{m \times (n+1)}$ the matrix containing the corresponding signals from n local receptors. Assuming a linear wide-field integration unit with weight vector $\mathbf{w} \in \Re^{n+1}$, the receptive field response is

$$\mathbf{r} = \mathbf{Sw} \, . \tag{4.2}$$

In general, $\mathbf{S}$ cannot be inverted directly since $m \neq n + 1$, so in order to obtain the weight vector $\mathbf{w}$ the Moore-Penrose inverse $\mathbf{S}^+$ must be used instead and gives an optimal solution

$$\mathbf{w} = \mathbf{S}^+ \mathbf{r} \tag{4.3}$$

in the following sense (Press, Teukolsky, Vetterling, and Flannery, 1992):

- If $\mathbf{S}$ can be inverted then $\mathbf{S}^{-1} = \mathbf{S}^+$ and thus $\mathbf{w} = \mathbf{S}^{-1}\mathbf{r}$.

- If there is more than one solution to Eq. (4.2) then Eq. (4.3) yields the solution with minimal norm $\|\mathbf{w}\|$.

- If Eq. (4.2) has no solution then the optimal weight vector $\mathbf{w}$ is determined in the sense of least squares with $\|\mathbf{Sw} - \mathbf{r}\| = \min$.

A proof is given, e.g., by Blanz (2000). In this study $m > n + 1$, thus we obtain a least squares solution.

The Moore-Penrose inverse can be efficiently calculated using a Singular Value Decomposition (SVD). The SVD of a matrix $\mathbf{M} \in \Re^{k \times l}$ yields three matrices $\mathbf{U}$, Σ and $\mathbf{V}$ with

$$\mathbf{M} = \mathbf{U}\Sigma\mathbf{V}^T \, . \tag{4.4}$$

These matrices have the following properties (Blanz, 2000):

- $\mathbf{U} \in \Re^{k \times l}$ is orthogonal in the columns: $\mathbf{U}^T\mathbf{U} = id_l$,

- $\Sigma \in \Re^{l \times l}$ is a diagonal matrix containing the singular values $\sigma_i \geq 0$, and

- $\mathbf{V} \in \Re^{l \times l}$ is orthogonal in the rows and columns: $\mathbf{V}^T\mathbf{V} = \mathbf{V}\mathbf{V}^T = id_l$.

Press et al. (1992) describe an algorithm for the numerical computation of the SVD.

Applied to the data matrix $S \in \Re^{m \times (n+1)}$ the SVD becomes

$$S = U\Sigma V^T \qquad (4.5)$$

with $U \in \Re^{m \times (n+1)}$, $\Sigma \in \Re^{(n+1) \times (n+1)}$ and $V \in \Re^{(n+1) \times (n+1)}$.

The inverse

$$\tilde{\Sigma} = [\text{diag}(\tilde{\sigma}_i)] \qquad (4.6)$$

of the diagonal matrix Σ is determined by inverting the diagonal elements

$$\tilde{\sigma}_i = \begin{cases} 1/\sigma_i, & \sigma_i > \theta \\ 0, & else \end{cases} . \qquad (4.7)$$

Singular values below a threshold θ are set to zero. Since U and V are orthogonal they can be inverted by transposing, and we yield the Moore-Penrose inverse

$$S^+ = V\tilde{\Sigma}U^T \qquad (4.8)$$

and the optimal weight vector

$$w = V\tilde{\Sigma}U^T r . \qquad (4.9)$$

However, the high dimensionality of the data matrix S impedes the numerical tractability of the SVD. As an example, for altitude control the output signals of 1920 elementary motion detectors are recorded at 1000 randomly selected locations in the virtual environment, each measurement for 10 time steps with a randomly chosen altitude, resulting in a very large data matrix $S \in \Re^{10000 \times 1921}$.

The dimensionality can be reduced by recording $S^T r \in \Re^{n+1}$ instead of $r \in \Re^m$ and $S^T S \in \Re^{(n+1) \times (n+1)}$ instead of $S \in \Re^{m \times (n+1)}$. Eq. (4.2) then becomes

$$S^T r = S^T S w , \qquad (4.10)$$

and with the Moore-Penrose inverse $\left(S^T S\right)^+$ we obtain

$$\left(S^T S\right)^+ S^T r = \left(S^T S\right)^+ S^T S w , \qquad (4.11)$$

hence the weight vector is

$$w = \left(S^T S\right)^+ S^T r . \qquad (4.12)$$

Again we use the SVD

$$S^T S = U\Sigma V^T \qquad (4.13)$$

to yield the Moore-Penrose inverse

$$\left(S^T S\right)^+ = V\tilde{\Sigma}U^T \qquad (4.14)$$

and finally the optimal weight vector

$$\mathbf{w} = \mathbf{V}\tilde{\Sigma}\mathbf{U}^\mathrm{T}\mathbf{S}^\mathrm{T}\mathbf{r}\,. \tag{4.15}$$

In principle, this method allows to record time series of arbitrary length since $\mathbf{S}^\mathrm{T}\mathbf{r}$ and $\mathbf{S}^\mathrm{T}\mathbf{S}$ remain at a constant size. In practice, however, the numerical stability of the SVD depends on the 'condition number' (Press et al., 1992), the ratio of the largest and the smallest singular value occurring in Σ. The condition number rapidly increases with an increasing number of measurements m, which could lead to errors for $m \gg n$.

4.3   Simulation Setup and Implementation

All experiments were implemented as computer simulations on a standard PC equipped with 3D OpenGL graphics acceleration for polygon-based, textured rendering of highly detailed scenes. The almost photorealistic visual stimuli generated by this system were used for both the receptive field optimization and the subsequent open-loop and closed-loop evaluation of the control strategies.

4.3.1   Virtual Environment

Fig. 4.2 shows two example views of the virtual environment used in the simulation experiments. The simulated environment is composed of a realistic, three-dimensional landscape with an uneven, hilly terrain and randomly placed trees. All visible objects such as terrain, obstacles and sky exhibit a detailed, high-resolution and nonperiodic surface texture with local variations in luminance, color, contrast and spatial frequency. The relative scale of the objects is chosen to match that of real-world objects. In the following, the units meter [m] and second [s] refer to simulated space and time, respectively. They are used to facilitate a comparison of the simulation results with potential real-world robot implementations of the proposed control strategies. The visual input is generated and processed with a temporal resolution of 50 updates per second of simulated time.

4.3.2   Omnidirectional Eye Model

Visual stimuli are retrieved from the virtual environment by an insect-inspired, omnidirectional eye model. The model incorporates typical sampling and filtering properties of insect compound eyes, such as an omnidirectional distribution of discrete light receptors in a spherical field of view, low image resolution, and overlapping receptive fields of the individual receptor units. Fig. 4.3a shows a quasi-homogeneous distribution of 642 local viewing directions on the sphere, arranged on the hexagonal lattice

Figure 4.2: Photorealistic virtual environment used in the simulation experiments. The example scenes show a realistic landscape with an uneven, hilly terrain and randomly placed trees. All visible objects such as terrain, obstacles and sky have a detailed, high-resolution and nonperiodic surface texture with local variations in luminance, color, contrast and spatial frequency.

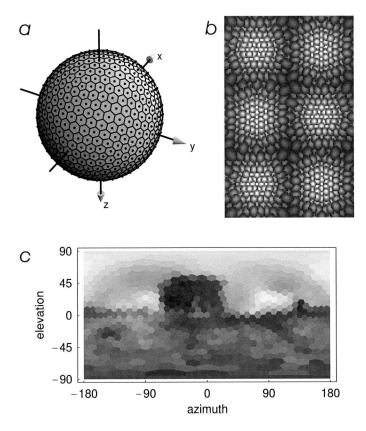

Figure 4.3: Insect-inspired omnidirectional eye model. *a*. Quasi-homogeneous distribution of 642 local viewing directions on the sphere, arranged as a hexagonal lattice. *b*. The spherical image is downsampled from six perspective views of the environment, taken at the current eye position. All local receptor units have appropriately distorted, overlapping Gaussian-shaped sensitivity distributions to prevent spatial aliasing. *c*. Virtual landscape as seen through the spherical eye. The image encompasses the entire sphere surrounding the current eye position, ranging from $-180°$ to $180°$ azimuth and from $-90°$ to $90°$ elevation. The increasing distortions toward the poles are due to the Mercator projection used for visualization and are not present in the original spherical image.

of an icosahedral geodesic grid of resolution level $R = 3$ (cf. Chapter 3). The average inter-receptor angle is $\Delta\varphi = 8.6°$. For comparison, each eye of the fruit fly *Drosophila* has approximately 700 ommatidia with an average inter-ommatidial angle $\Delta\varphi = 5°$ (Land, 1997b). Using this eye model, low-resolution, spherical images are generated by downsampling from multiple perspective views of the environment, taken at the current eye position. To reduce spatial aliasing, all local receptor units have appropriately distorted, overlapping Gaussian-shaped sensitivity distributions with an acceptance angle of half-width $\Delta\rho = \Delta\varphi = 8.6°$ (Fig. 4.3*b*). Many insect eyes sub-sample the environment with a reduced ratio of $\Delta\rho/\Delta\varphi \approx 1$ compared to the theoretically optimal value of $\Delta\rho/\Delta\varphi = 2$, presumably to increase image contrast (Land, 1997b). Fig. 4.3*c* shows the virtual landscape as seen through the spherical eye. The image covers the entire sphere surrounding the current eye position, ranging from $-180°$ to $180°$ azimuth and from $-90°$ to $90°$ elevation. The increasing distortions toward the poles are due to the Mercator projection used for visualization and do not occur in the original spherical image. For a detailed description of the compound eye simulation model see Chapter 2.

4.3.3 Spherical Image Processing

In analogy to the insect visual system, the spherical views acquired through the eye model are subject to further image processing in subsequent, retinotopic layers (cf. Section 3.1.1). In this simulation, three local operations are applied to each location in the spherical input images: contrast enhancement by spatial high pass filtering, time delay by temporal low pass filtering, and local motion detection using a correlation-type EMD model. All operations are performed in a retinotopic manner on a spherical geodesic grid of resolution $R = 3$. See Chapter 3 for implementation details on geodesic grid-based spherical image processing.

Contrast Enhancement

Each spherical input image is spatially high pass filtered by convolution with a DOG (Difference of Gaussians) kernel represented by a filter mask with a diameter of three pixels. This operation removes the local average intensity from each location in the image and thereby enhances the contrast, i.e., intensity differences between adjacent pixels (see Section 3.4 for details). The corresponding hexagonal and pentagonal filter kernels are depicted in Fig. 3.9*c* and *f*.

Local Motion Detection

In a subsequent processing step, the spherical distribution of local image motion is calculated from the incoming, high pass filtered intensity signals. 1920 insect-inspired

EMDs (elementary motion detectors) of the correlation-type after Hassenstein and Reichardt (1956) are placed between adjacent pixels of the spherical sampling grid. This results in an additional, geodesic grid-based spherical image containing one local motion signal for each edge (Section 3.6.2).

Fig. 3.11 illustrates the principle of the EMD type employed in this study. Two neighboring, but spatially distinct sampling points receive the input signals I_1 and I_2. In each of two mirror-symmetric semidetectors one of the input signals is temporally delayed and subsequently correlated with the other, undelayed signal. As suggested in Section 3.6.2, the delay stage of each EMD is implemented as a standard IIR (infinite impulse response) temporal low pass filter

$$L(I_t) = (1 - \frac{1}{\tau}) \cdot L(I_{t-1}) + \frac{1}{\tau} \cdot I_t, \quad \tau \geq 1. \tag{4.16}$$

Here, a time constant of $\tau = 5$ time steps is used for the temporal delay, which corresponds to 100 ms of simulated time. Correlating the delayed and undelayed input signals from opposite sides and subtracting both semidetector responses yields the full EMD output signal

$$\Phi(I_1, I_2) = L(I_1) \cdot I_2 - I_1 \cdot L(I_2). \tag{4.17}$$

Note that the response of a single EMD depends on the spatial structure and contrast of the input pattern. Moreover, the response signal is ambiguous and does not represent true image velocity. It is maximal when the image displacement during the temporal delay equals the angular distance between the two sampling points, and smaller for both faster and slower image motion. However, the sign of the average response signal reliably indicates the direction of motion, and for small image velocities the absolute value of the detector response increases almost linearly. The properties of EMDs are comprehensively discussed, e.g., by Borst and Egelhaaf (1989).

4.3.4   Flight Control Loop

The control architecture of the flying agent includes multiple processing stages and is depicted in Fig. 4.4. The spherical images composed of local light intensities and local optic flow provide the input signals for the wide-field integration units in the subsequent processing step. These units are tuned to specific patterns in the input signal distribution which are immediately relevant for the control of specific behaviors. In particular, they are optimized to estimate the roll and pitch angles directly from the omnidirectional intensity distribution, as well as yaw rotation and the relative nearness of objects in the frontolateral and ventral visual field from the local optic flow signals. The exact sensitivity distributions resulting from the optimization procedure described above are presented in the following section.

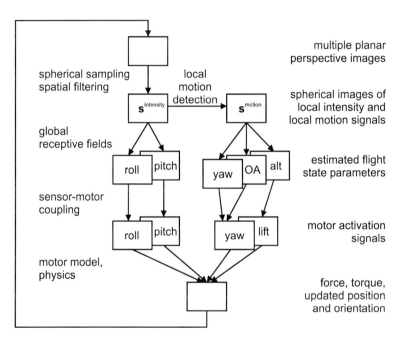

Figure 4.4: Image processing and flight control loop. Arrows represent processing stages, rectangles the output data generated by each stage. A low resolution spherical image of local light intensities is generated according to the eye model. A second spherical image containing local optic flow signals is generated from the intensity image by an array of elementary local motion detectors. Both images are analyzed by global receptive fields highly specialized on particular patterns in the input population which correlate with specific flight state parameters. Deviations from the desired flight state are detected and used to generate compensatory motor activations. The control loop is closed by updating the position and orientation of the agent.

The output values r_{roll}, r_{pitch}, r_{yaw}, r_{obst} and r_{alt} of the wide-field integration units (cf. Eq. 4.1) are immediately used to modulate motor activations that compensate for deviations from the desired flight state. The motor system of the simulated flying agent is capable of generating torque about the roll, pitch and yaw axes, as well as a lift force along the vertical body axis. The corresponding motor activation signals are m_{roll}, m_{pitch}, m_{yaw} and m_{lift}. At each simulation step, the motor activation vector

$$\mathbf{m} = (m_{\text{roll}}, m_{\text{pitch}}, m_{\text{yaw}}, m_{\text{lift}})^{\text{T}} \tag{4.18}$$

is updated by the vector of current response signals of the wide-field integration units

$$\mathbf{s} = (r_{\text{roll}}, r_{\text{pitch}}, r_{\text{yaw}}, r_{\text{obst}}, r_{\text{alt}})^{\text{T}} \tag{4.19}$$

using a linear sensor-motor transformation

$$\mathbf{m} = \mathbf{W}_{\text{sm}} \mathbf{s} \tag{4.20}$$

with the sensor-motor connection weight matrix

$$\mathbf{W}_{\text{sm}} = \begin{pmatrix} -w_{\text{roll}} & 0 & 0 & 0 & 0 \\ 0 & -w_{\text{pitch}} & 0 & 0 & 0 \\ 0 & 0 & -w_{\text{OMR}} & w_{\text{OA}} & 0 \\ 0 & 0 & 0 & 0 & w_{\text{lift}} \end{pmatrix} . \tag{4.21}$$

Inertial properties of the flying agent are ignored in this simulation. Small flying insects experience strong viscous air resistance and reach a steady state velocity after a short initial acceleration phase (Nachtigall, 1968). Thus to a first approximation, the velocity of the agent can be set proportional to the generated lift and thrust forces. The control loop is closed by updating the agent's position and orientation in the environment.

4.4   Results

The control architecture of the flying agent is composed of four independent behavioral modules, including attitude control, course stabilization, obstacle avoidance and altitude control. The central processing element for each of these sub-behaviors is a highly specialized global integration unit which extracts the required information from the spherical distribution of local input signals. This section shows the sensitivity distribution of each receptive field resulting from the optimization method described above, as well as the open-loop tuning curves of each unit. Further, examples for closed-loop flight behavior based on these units are presented.

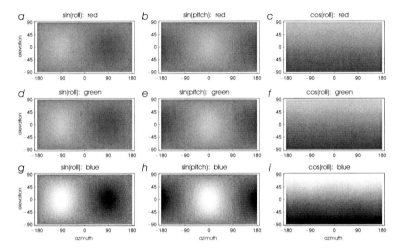

Figure 4.5: Optimal spherical receptive fields for attitude estimation from the brightness and color distribution around the current eye position. The receptive fields are tuned to the sines of the roll (*left column*) and pitch (*center column*) angles as well as the cosine of the roll angle (*right column*). The local weights for the red, green and blue components are depicted as separate gray value images. Dark regions indicate negative, bright regions positive weights, respectively.

4.4.1   Optimal Receptive Fields

For each target behavior, 10000 complete, spherical images of local intensity and local motion signals were recorded at randomly chosen locations in the virtual environment. The corresponding motor activations were varied within the limits that typically occur under free flight conditions, and recorded together with the simultaneously perceived visual stimulus. From these data, optimal linear receptive fields were determined using the optimization procedure described in Section 4.2. The resulting sensitivity distributions are presented in the following.

Attitude Angles

In this simulation the dynamic range of luminance is smaller than in the real world. To compensate, vertical color gradients are exploited instead of a luminance gradient. Fig. 4.5 shows optimal spherical receptive fields for attitude estimation from the momentary color distribution around the agent. The highest sensitivity is assigned to the

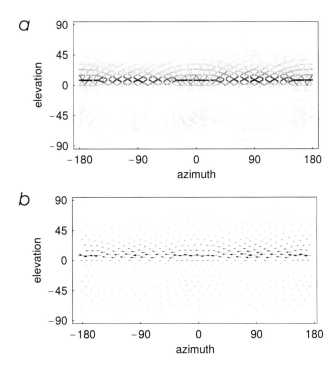

Figure 4.6: Optimal receptive field for the estimation of yaw rotations from the distribution of local image motion around the agent. *a*. Spherical distribution of local weights for all 1920 elementary motion detectors (EMDs). The orientation of each line indicates the local EMD direction, the gray level indicates the scaled absolute value of the corresponding weight, with darker lines representing higher values. *b*. Corresponding optic flow field. The orientation of each arrow indicates the local preferred direction of the receptive field, and the length denotes the local relative sensitivity.

blue component since the sky color is almost position invariant and therefore indicates the world vertical more reliably than the variable and position-dependent ground colors. The receptive fields are optimized to estimate the sines of the roll (left column) and pitch (center column) angles as well as the cosine of the roll angle (right column). As expected, the maximum sensitivity of the sine-of-roll sensor is located at +90° and -90° azimuth and 0° elevation. The sensitivity distribution of the sine-of-pitch sensor is shifted by 90° in azimuth direction relative to the roll sensor. The weights of the cosine-of-roll sensor are maximal at +90° and -90° elevation, therefore a cosine-of-pitch sensor (not shown) would have identical directions of maximum sensitivity. The output values of the sensors tuned to the sines of roll and pitch indicate the direction of deviation from a neutral attitude and can be used immediately as negative feedback signals for attitude control.

The sensitivity distributions of the receptive fields obtained in this study resemble those of insect ocelli. The ocelli are additional light sensors located on the dorsal side of the head. They integrate over large portions of the viewing sphere and are involved in the dorsal light response of many insects (Schuppe and Hengstenberg, 1993).

Yaw Rotation

An optimal receptive field for the detection of yaw rotations in the presence of simultaneous translatory forward motion is presented in Fig. 4.6 which shows the spherical weight distribution for the local EMD signals. The receptive field is tuned to rotations about the vertical axis during forward flight. The sensitivity for local image motion is highest in the region around the horizon, where the optic flow is largely independent of distance variations and translatory self-motion. The ventral region is exposed to strong translatory flow due to the smaller distance to the ground. In the dorsal region the insufficient contrast of the sky texture prevents a reliable motion detection. Thus, both regions do not correlate with yaw rotation and receive smaller weights. The output signal of this receptive field can be transmitted directly to the yaw motor to establish a basic course stabilization behavior resembling the optomotor response in insects.

Relative Obstacle Nearness

Optic flow induced by translatory self-motion contains information about the spatial layout of the environment as it depends not only on the velocity of the observer, but also on the distance of the observed objects, i.e., objects close to the observer induce stronger relative image motion than distant objects. If the translatory velocity is known, the optic flow is proportional to the reciprocal value of the absolute object distance, i.e., the absolute nearness of an object. For unknown velocities the nearness can be determined up to a scaling factor. The receptive field depicted in Fig. 4.7 is tuned to detect differences between the translatory optic flow in the left and right frontolateral

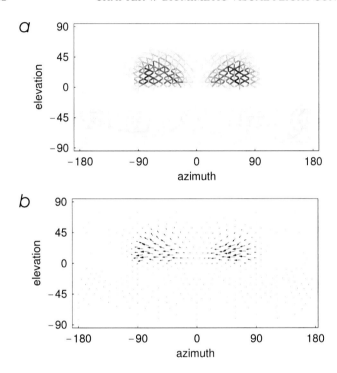

Figure 4.7: Optimal receptive field for the estimation of the frontolateral relative nearness difference from the distribution of local image motion around the agent. *a*. Spherical distribution of local weights for all 1920 elementary motion detectors. *b*. Corresponding optic flow field. Details as in Fig. 4.6.

regions of the visual field. The sign of the output signal indicates whether the relative object nearness is larger on the left or on the right side of the agent. It can be used to avoid potential obstacles by turning towards the direction of lesser image motion.

Relative Ground Nearness

The receptive field shown in Fig. 4.8 provides a further example for visual range finding from translatory optic flow. It is specialized on estimating the relative nearness of the agent to the terrain surface during forward flight. Naturally this information is

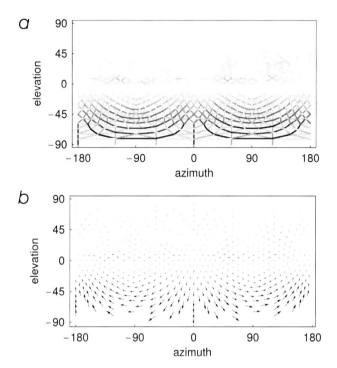

Figure 4.8: Optimal receptive field for the estimation of the relative ground nearness from the distribution of local image motion around the agent. *a*. Spherical distribution of local weights for all 1920 elementary motion detectors. *b*. Corresponding optic flow field. Details as in Fig. 4.6.

essential for altitude control and can be readily used to modulate the lift force. The receptive field exhibits a strong vertical anisotropy reflecting invariant properties of the environment and the typical behavior of the agent. The highest correlation of the EMD signals with the relative ground nearness occurs in the ventral region of the visual field since the ground plane provides a high contrast texture and a small absolute distance from the agent. Therefore, the largest weights are assigned to the ventral region. In contrast, the large distances and low contrast in the dorsal region render the optic flow largely independent of distance variations. Thus, image motion in the dorsal region does not predict ground nearness and therefore receives small weights.

4.4.2   Open-Loop Tuning Curves

Each receptive field is tuned to a particular class of input patterns which correlate with a specific behavioral component. Thus, the response of the corresponding wide-field integration unit is expected to be strongest when the current input pattern matches the receptive field layout. Each unit is designed to respond to its specific target behavior in a highly selective manner while showing the least possible sensitivity to other behaviors occurring simultaneously. In addition, the receptive field model allows, within a certain range, to quantitatively estimate the target parameters such as the rotatory velocity or the nearness to the ground plane or to an obstacle.

In this section, the open-loop tuning curves of all receptive fields are presented, and their implications for closed-loop behavior are discussed. For each measurement, the agent was passively moved in the virtual environment, using the same conditions and ranges of parameters that occur during typical flight behavior.

Attitude Angles

The open-loop tuning curves of the attitude sensors (Fig. 4.5) are shown in Fig. 4.9. As expected, the output signals of the roll and pitch sensors are not independent from each other, i.e., the roll signal is influenced by the pitch angle, and conversely. However, the simplest form of an attitude control mechanism requires only the signs of the attitude angles since they indicate the direction in which the agent needs to be rotated in order to restore a neutral attitude parallel to the ground plane. Fig. 4.9*a* and *d* demonstrate that the sign of one attitude angle can always be determined independently from the second angle. The only exceptions are two polar singularities occurring when the complementary angle reaches $\pm 90°$. In this case, however, the complementary receptive field yields its maximum response. This ensures that for any given combination of attitude angles at least one sensor is operational. Thus, the polar singularities for one sensor are fully compensated by the second sensor, and in a closed control loop the agent always has sufficient information to return to a neutral attitude. For example, the response of the roll sensor (Fig. 4.9*a*) is indifferently zero for all roll angles when the pitch angle is $\pm 90°$. In this situation, however, the pitch sensor is maximally activated (Fig. 4.9*d*). In a closed control loop this evokes a compensatory rotation about the pitch axis and thereby terminates the singularity for the roll sensor.

There are only two situations in which both the roll and the pitch signals are simultaneously zero and the system is at equilibrium. The equilibrium is stable when the system has reached a neutral attitude, and unstable when the agent is oriented exactly parallel to the ground plane, but with the ventral side pointing upwards. In the latter case, minimal changes of one attitude angle, as well as minimal anisotropies in the brightness or color distribution around the agent (which are always present in natural environments) suffice to trigger rotations away from the unstable and towards the

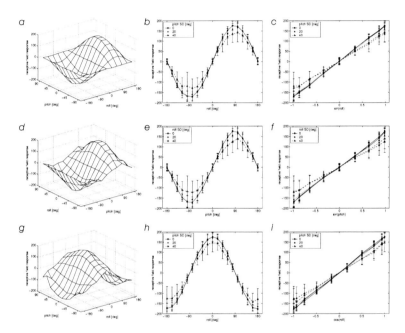

Figure 4.9: Open-loop tuning curves of three different attitude sensors, tuned to the sines of the roll (*a,b,c*) and pitch (*d,e,f*) angles, and to the cosine of the roll angle (*g,h,i*). The responses are averaged over *N*=50 measurements at randomly selected locations in the virtual environment. The left column depicts the mean responses as functions of both roll and pitch angles. In the center column the mean responses (symbols) and standard deviations (error bars) are shown as functions of either the roll (*b,h*) or the pitch (*e*) angle. In each plot, the complementary attitude angle is chosen randomly with zero mean and three different standard deviations as indicated by different symbols (circles, asterisks, triangles). The right column depicts the mean responses and standard deviations as functions of the sine of roll (*c*), the sine of pitch (*f*) and the cosine of roll (*i*).

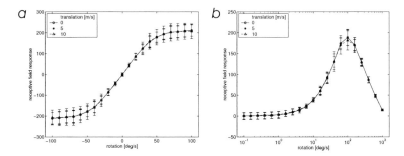

Figure 4.10: Open-loop tuning curve of the yaw sensor as a function of angular velocity (N=100). In each plot three different measurements for different translatory velocities are shown as mean values (symbols) with standard deviations (error bars). The similarity of the curves indicates that the response is robust against simultaneous translatory motion. *a*. For small angular velocities the response is approximately linear. *b*. The correct direction of motion is reliably detected for a velocity range of at least three orders of magnitude.

stable equilibrium.

Due to the system's tendency towards a neutral attitude it is unlikely that all possible combinations of roll and pitch angles occur during typical flight behavior. Rather, both angles remain close to zero. To reflect this situation the tuning curves of the roll and pitch sensors were measured separately as functions of one attitude angle while the complementary attitude angle was varied randomly with different standard deviations (Fig. 4.9*b,e*). The data indicate that even for a large standard deviation of $\pm 40°$, the correct signs of both attitude angles can be reliably detected. Furthermore, the tuning curves closely resemble the sines of the roll and pitch angles, giving additional, quantitative information (Fig. 4.9*c,f*).

Yaw Rotation

The yaw sensor (Fig. 4.6) is optimized to detect rotatory self-motion about the vertical axis in the presence of translatory motion. Fig. 4.10 depicts the open-loop response as a function of rotatory velocity. The tuning curve exhibits typical properties of the correlation-type EMD used to determine local image motion. For small angular velocities the receptive field response is approximately linear (Fig. 4.10*a*) and can be used for self-motion estimation and visual path integration (Franz et al., 1999). Beyond the linear range the response becomes ambiguous since after reaching a maximum, the signal decays with increasing angular velocity (Fig. 4.10*b*). However, the sign of the response signal robustly indicates the correct direction of rotation over multiple

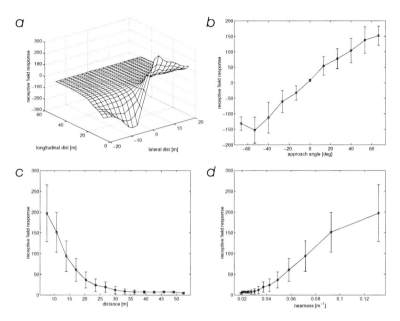

Figure 4.11: Open-loop response of the obstacle sensor. *a*. Mean receptive field response (*N*=100) during translatory forward motion ($v = 3\,m/s$) as a function of the longitudinal and lateral distance from an obstacle located at (0,0). *b,c,d*. Mean (circles) and standard deviation (error bars) of the response signal as functions of approach angle (*b*), obstacle distance (*c*), and obstacle nearness, i.e., reciprocal distance (*d*).

orders of magnitude. In a closed control loop this information is sufficient for a basic optomotor response and course stabilization mechanism.

The data also show that the response is robust against simultaneous translatory motion of the agent since the tuning curves are almost identical for different translatory velocities. This indicates a good de-coupling of rotation and translation by the specific sensitivity distribution of the optimal receptive field, and demonstrates the highly selective extraction of behaviorally relevant information from the visual input.

Relative Obstacle Nearness

The detection and avoidance of potential obstacles during forward translation requires information on both the distance and the direction of objects appearing in the frontal

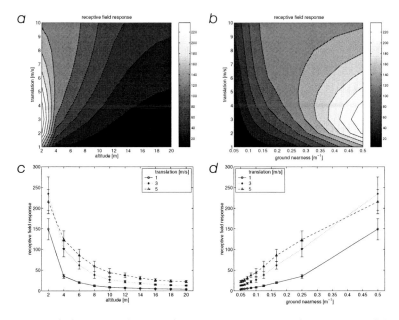

Figure 4.12: Open-loop tuning curve of the ground nearness sensor. *a,b*. Mean receptive field response (*N*=50) as a function of altitude and translatory velocity (*a*), and of ground nearness (i.e., reciprocal altitude) and translatory velocity (*b*). *c,d*. Mean (symbols) and standard deviation (error bars) of the receptive field response (*N*=100) for three selected translatory velocities as functions of altitude (*c*) and ground nearness (*d*).

and frontolateral field of view. Fig. 4.11*a* shows the open-loop response of the corresponding receptive field (Fig. 4.7) as a function of the longitudinal and lateral distance from a tree-shaped obstacle (Fig. 4.2*b*). The signal rapidly increases in the vicinity of the obstacle and can be used without further processing to trigger appropriate avoidance maneuvers. The sign of the response signal indicates on which side of the agent the obstacle is located and thereby determines which direction of rotation orients the agent towards open space.

Further analysis shows that the response signal contains quantitative information on both the relative direction (Fig. 4.11*b*) and distance (Fig. 4.11*d*) of obstacles. This information is implicitly used by the simple but effective obstacle avoidance strategy proposed above and needs not be computed explicitly. Thus, exploiting nearness instead of distance information obliterates the need for a division operation.

Relative Ground Nearness

The open-loop response of the ground nearness sensor (Fig. 4.8) is depicted in Fig. 4.12. As expected, the signal depends on both the altitude and the translatory velocity of the agent. It is highest when flying close to the terrain surface and decays rapidly with increasing altitude, therefore it indicates ground nearness, i.e., the reciprocal of altitude. This qualitative response characteristic holds for a wide range of translatory velocities (Fig. 4.12a,c). For velocities below $5\,m/s$ the signal increases almost linearly with ground nearness (Fig. 4.12b,d). Conversely, the reciprocal value of the response signal is proportional to the distance from the ground plane and could therefore be used to determine the true altitude of the agent. However, in a closed control loop the nearness signal is more useful than true altitude information since it can be used as a direct feedback signal to modulate the lift force. Thus, in analogy to obstacle avoidance, a division operation is not required for altitude control.

4.4.3 Closed-Loop Flight Behavior

In the preceding sections I derived optimal receptive fields for wide-field integration units tuned to specific, flight-relevant patterns in the spherical distribution of local light intensity and local image motion. The individual units are sensitive to attitude (i.e., roll and pitch), rotation about the yaw axis, relative obstacle nearness, and relative ground nearness, respectively. Their open-loop tuning curves show that the response of each unit is suitable as a direct feedback signal for the corresponding control mechanism without further processing.

Here I demonstrate this type of immediate, reactive control of spatial behavior from visual cues by closing the control loop according to Eq. 4.20. Four independent, visual orientation strategies for attitude control, course stabilization, obstacle avoidance and altitude control are combined in a flying autonomous agent and tested in closed-loop simulations. The resulting trajectories shown in this section demonstrate that the agent is capable of autonomous flight control with all six degrees of freedom in a realistic virtual environment.

Attitude Control

Attitude control ensures that the agent always maintains a neutral, horizontal orientation with respect to the terrain surface, i.e., it aligns the vertical axes of the body and world coordinate systems. This is a prerequisite for all remaining visual control strategies which analyze local image motion in specific viewing directions. The mechanism for visual attitude control is inspired by the dorsal light response in insects, and evaluates the spherical distribution of luminance and color in the perceived stimulus. In particular, it exploits the position-invariant, vertical intensity gradient in the environment.

Figure 4.13: Course stabilization. In the absence of potential obstacles yaw rotations are inhibited by the optomotor response, and therefore the agent moves on a straight path. The trajectories are shown as bright lines, together with their projections on the terrain surface. The position and orientation of the artificial agent is indicated by a small helicopter. All trajectories were recorded during a closed-loop autonomous flight with all six degrees of freedom.

Figure 4.14: Obstacle avoidance. Strong frontolateral image motion during forward translation of the agent indicates a potential obstacle. The agent avoids collisions by turning toward the direction of smaller optic flow indicating a larger relative object distance. Trajectory visualization as in Fig. 4.13.

Figure 4.15: Altitude control and terrain following. The agent follows the terrain elevation by modulating the lift force proportional to the average translatory image motion in the ventral field of view. Slight oscillations in altitude can be observed when flying over regions with strong changes in texture luminance or contrast. In all situations, ground collisions are robustly avoided. Trajectory visualization as in Fig. 4.13.

As shown in Fig. 4.9, the global receptive fields sensitive to roll and pitch (Fig. 4.5) reliably indicate the direction of deviation from neutral attitude independently of the local terrain texture or nearby obstacles. Thus, in the closed-loop simulation both the roll and pitch angles of the agent always remain close to zero.

Course Stabilization

Visual course stabilization closely resembles the well-known optomotor response in flies. Involuntary rotations with respect to the environment are inhibited by stabilizing the retinal image. Here, the agent detects large-field rotatory image motion in the presence of simultaneous translatory motion, using the optimized, spherical receptive field shown in Fig. 4.6. Since the response signal of this wide-field integration unit reliably indicates the direction of rotation over multiple orders of magnitude (Fig. 4.10), it provides a robust feedback signal for course stabilization. Fig. 4.13 shows two exemplary trajectories of the flying agent over the terrain surface, recorded during closed-loop autonomous flight with all six degrees of freedom. The examples demonstrate that in the absence of potential obstacles yaw rotations are inhibited by the optomotor response, and thus the agent moves on a straight path.

Obstacle Avoidance

Two exemplary obstacle avoidance maneuvers are depicted in Fig. 4.14. During forward translation of the agent, strong image motion in the left or right frontolateral region of the spherical field of view indicates a potential obstacle. Thus, obstacle avoidance behavior is controlled by a wide-field integration unit tuned to frontolateral translatory optic flow (Fig. 4.7). As illustrated in Fig. 4.11, the response signal of this unit contains information on both the relative nearness of an obstacle and the direction of approach. The agent avoids a collision by turning toward the direction of smaller optic flow which indicates a larger relative object distance.

Altitude Control and Terrain Following

The visual mechanism for altitude control is based on the optimized wide-field integration unit depicted in Fig. 4.8. During translatory forward motion of the agent, this unit extracts information on relative ground nearness from the optic flow in the ventral region of the spherical field of view (Fig. 4.12). The agent follows the terrain elevation by modulating the lift force proportional to the average ventral image motion. An "equilibrium altitude" is reached when the lift force equals gravity. Fig. 4.15 shows two exemplary trajectories demonstrating autonomous visual altitude control and terrain following behavior. Slight oscillations in altitude can be observed when flying over regions with strong changes in the luminance or contrast of the terrain texture. In all situations, collisions with the terrain surface are robustly avoided.

4.5   Discussion

In this chapter I presented a biologically motivated visual flight control system composed of four reactive orientation strategies including attitude control, course stabilization, altitude control and obstacle avoidance. Using a behavior-oriented approach, the entire vision system is designed to acquire and process only behaviorally relevant information in a highly selective manner. Since flying insects are optimized for flight behavior by natural evolution, they serve as exemplary systems for both the orientation strategies as well as the information acquisition and processing stages of the artificial vision system.

The orientation mechanisms are based on the evaluation of light intensity and optic flow patterns in a spherical field of view by highly specialized wide-field integration units inspired by fly tangential neurons. The covariance of local receptor signals and specific behaviors is used to optimize the sensitivity distributions of the global receptive fields to selectively extract flight-relevant information from the visual input. Explicit models of the distance or contrast distribution in the environment are not required since the weight distributions are derived directly from the input signals. The resulting optimal receptive fields as well as their open-loop tuning curves are presented and discussed for all four orientation strategies. Example trajectories from closed-loop flight simulations demonstrate that four independent, purely reactive visual orientation strategies are sufficient for robust, three-dimensional flight behavior in a highly realistic virtual environment.

4.5.1   Visual Flight Control

The visual orientation strategies described above are simple but highly effective. In the following I will discuss their advantages and limitations in the context of visual flight control in natural environments. In addition, biologically plausible enhancements are proposed for each mechanism, many of them requiring only minor modifications of the existing control strategies.

Attitude Control

The mechanism proposed for visual attitude control requires a position-invariant, vertical gradient of color or light intensity. This gradient is present in most natural, open environments, but may be reduced when the sky is partially occluded by large objects such as buildings or trees, or even completely absent inside a building. Evidently, insects are able to fly under these conditions using additional visual cues such as image motion and edge orientation, as well as other sensory modalities such as inertial cues and wing load (Hengstenberg, 1993).

The linear wide-field integration of light intensity makes the output signal sensitive to global changes in illumination as well as to local "outliers". In particular, direct sunlight might impede attitude sensing since it is by multiple orders of magnitude brighter than light coming from other directions. Although the sun occupies only a small solid angle in the viewing sphere, a simple optical or neural wide-field integration unit would be literally blinded as soon as any portion of the receptive field is exposed to direct sunlight. As above, insects are perfectly capable of flying under such conditions even though they cannot close or cover their eyes. A possible solution exploits the subdivision of compound eyes into discrete local receptor units. The signal of each unit can be normalized by an additional, nonlinear local processing element that is insensitive to the characteristic brightness and color of the terrain surface, but saturates when oriented towards the sky. This yields a binary spherical image serving as input for the subsequent wide-field integration units. Preliminary simulation experiments indicate that such local nonlinearities eliminate negative effects of direct sunlight and significantly increase the robustness of the dorsal light response (data not shown).

The response signals of both attitude sensors are simultaneously zero for neutral as well as for inverse attitude. The latter case, however, requires the strongest possible compensatory rotation. Thus, a further improvement of the attitude control strategy can be achieved by introducing an additional receptive field tuned to the cosine of the roll or pitch angle (Fig. 4.5, right column). The cosine sensor indicates the direction-invariant deviation from neutral attitude (Fig. 4.9g,h,i), whereas both sine sensors provide directional information. As the tuning curves of the attitude sensors closely resemble the sines of the roll and pitch angles (Fig. 4.9b,c,e,f) they could be used together with an additional cosine sensor to determine the true attitude angles. However, this would require inverse trigonometric operations of questionable biological plausibility, burdened with the problem of an appropriate normalization of the input signal. Considering the closed-loop experiments presented above, calculating the true attitude angles appears to be unnecessary since the immediate receptive field response is fully sufficient as a negative feedback signal for attitude stabilization.

Course Stabilization and Obstacle Avoidance

In this study, rotations about the yaw axis are simultaneously controlled by two antagonistic visual strategies, both evaluating optic flow patterns. The mechanism responsible for course stabilization inhibits yaw rotations by minimizing the perceived rotatory image motion. It resembles the well-known optomotor response in insects

(Reichardt, 1969; Götz, 1968) and stabilizes flight against involuntary deviations from a straight trajectory. In contrast, the obstacle avoidance mechanism elicits yaw rotations to prevent imminent collisions with objects in the environment. It compares the optic flow in the left and right frontolateral regions of the visual field and takes evasive action towards the direction of lesser image motion. This strategy exploits the fact that optic flow fields induced by translatory self-motion contain information about the three-dimensional structure of the environment (Koenderink and van Doorn, 1987) and can therefore be used for visual range finding (Srinivasan, 1993). Simultaneous rotatory self-motion needs to be suppressed since it corrupts the translatory flow field and impedes range finding. Thus, it appears to be a major function of course stabilization to ensure pure translatory optic flow.

However, rotations are inevitable to change the direction of flight, e.g., to avoid an obstacle. Therefore, they should be executed as fast as possible to interrupt translatory motion for the shortest possible time. It has been shown that flies use extremely fast, saccadic turns to change direction between periods of straight forward flight (Schilstra and van Hateren, 1999; Tammero and Dickinson, 2002). There is evidence that in the fruit fly *Drosophila melanogaster* saccades are triggered by frontolateral optic flow exceeding a threshold (Tammero and Dickinson, 2002). Thus, only minor modifications are required to enable saccades for the obstacle avoidance strategy proposed in this study. Further image stabilization can be achieved by compensatory head movements as observed in the blowfly *Calliphora erythrocephala* (Hengstenberg et al., 1986; van Hateren and Schilstra, 1999), thereby maximizing the periods of translatory optic flow between saccades.

Altitude Control

The altitude control mechanism proposed above is most effective for low altitude terrain following with the goal of avoiding ground contact. The robustness of this strategy is based on the steep, non-linear increase of ventral image motion at low altitudes (Fig. 4.12a,c). This response characteristic holds for arbitrary constant translatory velocities. In addition to vertical collision avoidance, the quantitative distance information contained in the response signal (Fig. 4.12b,d) can be used to control flight at arbitrary altitudes within a certain range. Both strategies require that the translatory velocity is controlled separately from altitude. In small flying insects this is facilitated by viscous air resistance, as pointed out in Section 4.3. Further sources of velocity information include the measurement of relative air flow and the visual tracking of objects or salient features of the environment. As an alternative, altitude and forward velocity can be simultaneously controlled by keeping the ventral optic flow constant. Experimental results by Srinivasan et al. (2000) suggest that this strategy is used by honeybees for landing on flat surfaces.

Visual range finding from image motion is problematic when flying at high altitudes. For large distances from the terrain surface the ventral optic flow induced by translatory motion is very small compared to the flow components elicited even by minimal rotations about the roll or pitch axes, making it very difficult to yield reliable distance information. Therefore, other cues such as the atmospheric pressure might be more helpful at high altitudes.

4.5.2 Applications

All four mechanisms for visual flight stabilization presented above are purely reactive and do not require internal representations of the environment or trajectory planning. They are based on a massively parallel, feed-forward flow of information with few sequential processing steps, leading to robust behavior and short reaction times suitable for real-time control of autonomous robots. The connection and weighting schemes are highly optimized for specific tasks and do not change during computation, facilitating small-scale, integrated hardware implementations. Analog VLSI (very large scale integration) technology appears particularly promising since first correlation-type EMDs have already been implemented (Harrison and Koch, 1999; Indiveri and Douglas, 2000) and used for a one-dimensional stabilization task (Harrison and Koch, 2000). This approach could be extended to more complex behavior by employing wide-field integration units as proposed in this work. Simple, robust control algorithms are crucial for autonomous vehicle guidance and robotics, especially in applications with strong constraints in size, weight and energy consumption, such as aerospace and miniature robotics.

However, real-world flying robots are subject to constraints in the available hardware platforms such as model planes, helicopters, or blimps. Most of these platforms are built at larger physical scales than flying insects. Therefore they may experience fundamentally different physical effects when interacting with the environment. In particular, the relative effect of viscous air resistance is strongly reduced in large scale robots compared to small flies. This has consequences for the applied orientation strategies and control algorithms, since large flying robots require a simultaneous, sensor-based control of both forward velocity and ground distance. In contrast, the forward velocity of small flying insects is to a first approximation proportional to the applied force. Furthermore, inertia needs to be actively compensated in large robots, whereas in small flies unintended translatory self-motion relative to the surrounding air is passively reduced by air resistance. Conversely, large robots are more robust against external influences like wind or turbulences, which require active compensation in insects.

Many flying insects are physically more robust and tolerant against collisions than robots, and are therefore able to survive failures of the obstacle avoidance system. For most flying robots collisions are fatal due to the large mass and the resulting forces during impact. In addition, even a slight contact of the propulsion system with an obstacle may lead to severe damage or destruction of the flight platform. Thus, large-scale flying robots require an extremely reliable mechanism for collision avoidance, whereas insects are inherently more tolerant against failures of the visual system. In conclusion, the described, insect-inspired reactive control mechanisms are expected to perform best for applications such as micromechanic implementations or small underwater robots, which are subject to similar physical effects and constraints as experienced by real flies. Larger robots may require different control algorithms due to different motion dynamics and propulsion systems. Conversely, miniature implementations or computer simulations may be more suitable models for flight control in insects than large-scale flying robots.

4.5.3   The Role of Learning

For all applications mentioned above the proposed behavior-oriented vision system provides a simple but highly effective method to extract flight-relevant information from the environment. In analogy to the simulation experiments described in this work, the basic processing architecture can be based on the insect visual system, whereas the exact sensitivity distributions of the receptive fields are derived from the covariance of motor activation and visual input during typical behavior in a typical environment, i.e., the sensor-motor mapping is acquired by learning. This has a number of advantages compared to other optimization methods based on explicit statistical models of the agent and the environment. Such models may be difficult to obtain, e.g., the distance distribution around a model helicopter during flight. In addition, all information needs to be explicitly incorporated into the model, thus potentially important aspects of the stimulus might be omitted, whereas other information contained in the model might not be required for the actual control task. In contrast, learning the sensor-motor mapping directly from the perceived stimulus and the concurrent motor activation inherently includes all relevant properties of the agent, the environment and the behavior. Further, the system is self-calibrating since random motor activations are sufficient to elicit different stimulus patterns. The sensor-motor mapping is determined from the covariance of these patterns with the original motor commands. This resembles the way animals and humans acquire strategies for sensor-motor coordination and movement control (Gibson, 1950).

In this study, linear receptive fields and a linear optimization method are fully sufficient for the complex control task of three-dimensional flight stabilization. Using nonlinear learning methods such as kernel regression (Schölkopf and Smola, 2002) may further improve the performance of the system since the mapping between sensory input and motor activation might be represented more accurately in a nonlinear space. However, linear receptive fields have the advantage of a simple and efficient implementation, requiring only one scalar product of the input and weight vectors. In addition, linear integration units serve as models for tangential neurons in flies (Krapp and Hengstenberg, 1996). It is worth noting that the input images containing pre-processed local information such as luminance, color, or local motion, may differ in size, shape or resolution, or may be extended to additional local input signals such as line orientation, color contrast or a different type of elementary motion detector. Moreover, the behavior-oriented approach is not restricted to vision. Other sensory modalities such as inertial sensors may be included in the optimization process. The exact response and noise characteristics of each sensor needs not be known. Learning ensures that the maximum amount of behaviorally relevant information is extracted even from arbitrary receptor configurations.

Chapter 5

Concluding Remarks

5.1 Summary

In the preceding chapters I presented a complete, biomimetic vision system with a spherical field of view. I introduced various novel algorithms and data structures for the efficient acquisition, representation, processing and behavior-oriented evaluation of spherical images in the context of visual flight control. All algorithms were designed in close resemblance to the different processing stages in the visual system of flying insects, and were implemented and tested in computer simulations.

The first processing stage of any vision system is the acquisition of images from the environment. The specific sampling grid of an eye or a camera, as well as the filtering properties of the light receptors determine the type and amount of information that is available for the control of a particular behavior. Conversely, the control of a specific behavior may benefit from a specific eye design. The compound eyes of flying insects are highly optimized for the acquisition of flight-relevant information from the environment: The spherical field of view facilitates the extraction of self-motion parameters from global optic flow patterns, the low image resolution minimizes the amount of parallel processing required for local motion detection, and the overlapping, Gaussian-shaped receptive fields of the individual receptor units prevent information which cannot be correctly sampled from entering the visual system. In Chapter 2, I presented an accurate and efficient simulation model for insect compound eyes. It employs hardware accelerated computer graphics and can therefore be used in complex and realistic virtual environments to reconstruct the visual stimuli perceived by a real insect during flight, and to acquire spherical input images in simulations of biomimetic artificial systems for visual flight control.

Image processing in the visual system of flying insects is highly specialized for the extraction of flight-relevant information from the perceived stimulus. Most of this information can be recovered from the spherical distribution of local optic flow. Thus, the predominant tasks of the first processing stages are spatiotemporal signal enhancement and local motion detection at every location in the spherical field of view. The insect brain performs the necessary operations such as local averaging, lateral inhibition and local correlation, in a massively parallel fashion. Its regular, retinotopic structure preserves local neighborhood relations in the spherical field of view throughout several successive layers. In Chapter 3, I transferred the design principles of the insect visual system to spherical machine vision. Sampling spherical images at the vertices of an icosahedral geodesic grid results in a spherical raster of discrete, hexagonal pixels and allows homogeneous, seamless image processing over the entire sphere. Moreover, the edges of a geodesic grid provide a homogeneous and isotropic sampling lattice for spherical flow fields, and can be used to implement spherical arrays of correlation-type elementary motion detectors in a direct and elegant manner. Possible applications are the reconstruction of neural activity patterns occurring in the first, retinotopic layers of the insect visual system during an experiment, as well as the efficient and homogeneous representation and processing of spherical images in computer vision, especially in the context of three-dimensional self-motion detection and flight control.

Basic stabilization behaviors of flying insects, such as the dorsal light response and the optomotor response, are stereotyped and almost purely reactive, but extremely fast. This indicates that only few sequential processing steps are involved between image acquisition and motor activation. In Chapter 4, I demonstrated that wide-field integration units inspired by fly tangential neurons allow to recover flight-relevant parameters in a highly selective manner directly from the spherical distribution of local intensity and local motion signals provided by the early, retinotopic processing layers. These units have complex receptive fields containing prior knowledge on invariant properties of both the environment and the behavior. I showed that an optimal sensitivity distribution can be derived directly from the correlation of the local input signals with a specific behavior, and presented optimal receptive fields for the estimation of attitude, rotation about the vertical axis, the relative nearness of obstacles, and the relative nearness of the terrain surface. In addition, I demonstrated that these receptive fields provide sufficient information for three-dimensional, visual flight control with all six degrees of freedom, including obstacle avoidance and terrain following in a realistic virtual environment.

5.2   Perspectives

Natural organisms and autonomous artificial systems interact with the environment in a closed loop of action and perception. By investigating visual information acquisition and processing, this dissertation focuses on the *perception* aspect of the control cycle. However, it is a valid assumption that in natural organisms the *action* aspect of the action-perception cycle, i..e., the actuator or motor system as well as the physical interaction with the environment, is optimized to a similar degree for the generation and control of specific behaviors. For instance, it may exploit physical effects such as the resonance of the flight motor system or self-generated vorticity in the air. Building biomimetic models of action-related components of an organism may - in analogy to the perception aspect described above - lead to the development of novel, efficient propulsion and motor control structures in artificial systems, which in turn can be used as models for the corresponding components of biological organisms. Most importantly, however, physical interaction with the environment has an immediate impact on the information processing and control strategies required to generate behavior. It determines how the body moves in reaction to particular motor commands. For instance, the same motor commands may produce different behaviors for different densities and viscosities of the surrounding medium. Thus, the next logical step towards a realistic simulation model of visually controlled behavior in insects is to develop an accurate model of the propulsion system and its physical interaction with the environment.

Appendix A

Tetrahedral Image Acquisition

The compound eye simulation model described in Chapter 2 generates low resolution spherical images from multiple planar perspective views. These planar views need to be arranged in a manner such that the entire spherical field of view is covered and the eye point is completely enclosed. In Chapter 2 this is achieved by using six square images arranged as a cube environment map (Fig. 2.6). However, the minimum number of planes required to enclose a point in 3D space is four. Thus, the entire viewing sphere can alternatively be represented by four planar perspective images, e.g., using a tetrahedron arrangement.

Fig. A.1 shows a tetrahedron environment map representing the same example scene as the cube map depicted in Fig. 2.6. Since both the cube and the tetrahedron environment map are composed of planar perspective images, the same sampling and filtering algorithms can be used for both configurations to obtain low resolution spherical images. In particular, changes in the receptive field projection method (Section 2.2.5) or in the lookup table representation of the distorted receptive fields (Section 2.2.7) are not required, and the resulting spherical images are identical. Fig. A.2 demonstrates how the receptive field projection method is applied to a tetrahedron environment map. As in cube map filtering, each receptor unit averages the light received in a conical region around its optical axis using a Gaussian-shaped sensitivity distribution. The receptive fields are projected onto the planar perspective images of the environment map by intersecting the viewing cones of the receptor units with the triangular faces of the tetrahedron environment map. The resulting, anisotropically distorted receptive fields are depicted in Fig. A.3.

Apparently the tetrahedron arrangement has the advantage that only four instead of six perspective images of the surrounding environment need to be generated each time the eye position is changed. However, in most practical applications, cube environment maps are superior to tetrahedron maps in terms of memory efficiency and filtering time. In the following it is shown that in cube maps both the input images and the distorted

125

Figure A.1: Example scene on a tetrahedron environment map. The environment map is shown unfolded (*upper right*) and on the surface of a tetrahedron (*lower left*). It is composed of four planar perspective images projected on equilateral triangles.

filter masks are smaller than in tetrahedron maps of equal image quality.

To yield an equal quality of anti-aliasing in the resulting spherical image both environment maps must have an equal minimum receptive field diameter Δa_{min} in the image plane for a given aperture angle $\Delta \alpha$ (Fig. 2.8). As pointed out in Section 2.2.6 the diameter Δa of the distorted receptive fields in the image plane is smallest when the receptor viewing direction is orthogonal to the corresponding face of the environment map. Thus, the appropriate relative sizes of the environment maps are determined by inscribing a sphere with radius r to both the cube and the tetrahedron (Fig. A.4a). Hence, the surface area of the cube is

$$A_{\text{cube}} = A_{\text{cubeEM}} = 24r^2 \qquad (A.1)$$

and equals the area of the corresponding cube environment map. The surface area of a tetrahedron with equal image quality is

$$A_{\text{tetra}} = 24\sqrt{3}r^2 \,, \qquad (A.2)$$

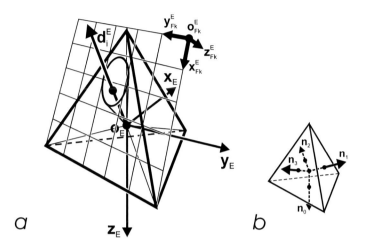

Figure A.2: Coordinate systems of a tetrahedron environment map. *a*. Eye (E) and face (F_k) coordinate systems. *b*. Normal vectors of the tetrahedron faces.

and thus the area to be filtered is larger by a factor of $\sqrt{3}$. In addition, a rectangular image needs to be generated for each triangular face of the tetrahedron map due to the operating principles of current hardware accelerated computer graphics. As shown in Fig. A.2*a* and Fig. A.4*b*, this doubles the effective area of the tetrahedron environment map to be rendered and transferred to CPU memory, and yields

$$A_{\text{tetraEM}} = 2A_{\text{tetra}} = 48\sqrt{3}r^2 \,. \tag{A.3}$$

Therefore, the area of a tetrahedron environment map exceeds the size of a cube map with equal image quality by a factor of $2\sqrt{3}$, although 50% of the rendered pixels are not used in the subsequent filtering stage.

A further aspect of tetrahedron maps is the need for larger and more distorted filter masks. The tetrahedron arrangement leads to larger eccentricity angles β between the receptor viewing directions and the normal vectors of the faces (cf. Fig. 2.8). This angle is maximal when the receptor axis is directed towards a corner of a face, resulting in $\beta_{\text{maxCube}} = 54.7°$ for the cube and $\beta_{\text{maxTetra}} = 70.5°$ for the tetrahedron. For instance, the projection of a circular receptive field with an aperture angle of $\Delta\alpha = 10°$ onto the center of a face (i.e., along the face normal vector) yields a circular filter mask with a normalized relative area of $A_{\text{RFcenter}} = 1.0$. However, projecting the same receptive field into a corner of an environment map face results in elliptical distorted

Figure A.3: Gaussian-shaped receptive fields projected on a tetrahedron environment map. For demonstration, large, non-overlapping receptive fields are used in this example.

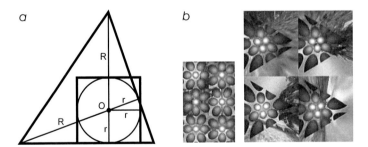

Figure A.4: Size comparison for cube and tetrahedron environment maps. *a*. Vertical section through a tetrahedron and a cube with a common center O and a common radius r of a sphere inscribed to both solids. R is the radius of the sphere circumscribed to the tetrahedron. *b*. Cube (*left*) and tetrahedron (*right*) environment maps with equal minimum receptive field diameter Δa_{min} in the image plane for a given aperture angle $\Delta\alpha$ (cf. Fig. 2.8 and Section 2.2.6).

filter masks with a relative area of $A_{\mathrm{RFcube}} = 5.32$ for the cube and $A_{\mathrm{RFtetra}} = 29.7$ for the tetrahedron. In this example, the same filter mask is larger on the tetrahedron map by a factor of 5.6.

In conclusion, cube environment maps should be preferred to generate spherical images since they are smaller than tetrahedron maps and require smaller filter masks. However, tetrahedron environment maps are useful in real world implementations of spherical vision using camera clusters. Separate cameras with different viewing directions are arranged to simultaneously record perspective images for the individual faces of the environment map (Swaminathan and Nayar, 2000). Here, a tetrahedron arrangement has the advantage that only four cameras and frame grabbers are required instead of the six units which are needed for the cube configuration.

Bibliography

Adelson, E. H., and Bergen, J. R. (1991). The plenoptic function and the elements of early vision. In Landy, M., and Movshon, J. A. (Eds.), *Computational Models of Visual Perception*, pp. 3–20. MIT Press, Cambridge, MA.

Augenbaum, J. M., and Peskin, C. S. (1985). On the construction of the Voronoi mesh on a sphere. *Journal of Computational Physics*, *59*, 177–192.

Barlow, H. B. (1952). The size of ommatidia in apposition eyes. *Journal of Experimental Biology*, *29*, 667–674.

Blanz, V. (2000). Automatische Rekonstruktion der dreidimensionalen Form von Gesichtern aus einem Einzelbild. Dissertation, Universität Tübingen, Germany.

Borst, A., and Bahde, S. (1988). Visual information processing in the fly's landing system. *Journal of Comparative Physiology A*, *163*, 167–173.

Borst, A., and Egelhaaf, M. (1989). Principles of visual motion detection. *Trends in Neurosciences*, *12*(8), 297–306.

Borst, A., and Egelhaaf, M. (1993). Detecting visual motion: Theory and models. In Miles, F. A., and Wallman, J. (Eds.), *Visual Motion and its Role in the Stabilization of Gaze*, pp. 3–27. Elsevier, Amsterdam.

Borst, A., and Haag, J. (2002). Neural networks in the cockpit of the fly. *Journal of Comparative Physiology A*, *188*, 419–437.

Braitenberg, V. (1984). *Vehicles: Experiments in Synthetic Psychology*. MIT Press, Cambridge, MA.

Buchner, E. (1971). Dunkelanregung des stationären Flugs der Fruchtfliege *Drosophila*. Diploma thesis, Universität Tübingen, Germany.

Buchner, E. (1976). Elementary movement detectors in an insect visual system. *Biological Cybernetics*, *24*, 85–101.

Bülthoff, H. H., Poggio, T., and Wehrhahn, C. (1980). 3-D analysis of the flight trajectories of flies (*Drosophila melanogaster*). *Zeitschrift für Naturforschung C - A Journal of Biosciences, 35c*(9/10), 811–815.

Burkhardt, D. (1989). Autrum's impact on compound eye research in insects. In Stavenga, D. G., and Hardie, R. C. (Eds.), *Facets of Vision*, pp. 15–29. Springer-Verlag, Berlin.

Burt, P. J., and Adelson, E. H. (1983). The Laplacian pyramid as a compact image code. *IEEE Transactions on Communications, COM-31*(4), 532–540.

Cartwright, B. A., and Collett, T. S. (1983). Landmark learning in bees. *Journal of Comparative Physiology A, 151*, 521–543.

Chahl, J. S., and Srinivasan, M. V. (1997). Reflective surfaces for panoramic imaging. *Applied Optics, 36*(31), 8275–8285.

Chahl, J. S., and Srinivasan, M. V. (2000). Filtering and processing of panoramic images obtained using a camera and a wide-angle-imaging reflective surface. *Journal of the Optical Society of America A, 17*(7), 1172–1176.

Cliff, D. (1991). The computational hoverfly: A study in computational neuroethology. In Meyer, J. A., and Wilson, S. W. (Eds.), *From Animals to Animats: Proceedings of the First International Conference on Simulation of Adaptive Behavior (SAB'90)*, pp. 87–96 Cambridge, MA. MIT Press Bradford Books.

Cliff, D. (1992). Neural networks for visual tracking in an artificial fly. In Varela, F., and Bourgine, P. (Eds.), *Towards a Practice for Autonomous Systems: Proceedings of the First European Conference on Artificial Life (ECAL'91)*, pp. 78–87 Cambridge, MA. MIT Press Bradford Books.

Collett, T. S. (1980a). Angular tracking and the optomotor response: An analysis of visual reflex interaction in a hoverfly. *Journal of Comparative Physiology A, 140*, 145–158.

Collett, T. S. (1980b). Some operating rules for the optomotor system of a hoverfly during voluntary flight. *Journal of Comparative Physiology A, 138*, 271–282.

Collett, T. S., and Land, M. F. (1975a). Visual control of flight behaviour in the hoverfly, *Syritta pipiens* L.. *Journal of Comparative Physiology, 99*, 1–66.

Collett, T. S., and Land, M. F. (1975b). Visual spatial memory in a hoverfly. *Journal of Comparative Physiology, 100*, 59–84.

Collett, T. S., Nalbach, H. O., and Wagner, H. (1993). Visual stabilization in arthropods. In Miles, F. A., and Wallman, J. (Eds.), *Visual Motion and its Role in the Stabilization of Gaze*, pp. 239–263. Elsevier, Amsterdam.

Dahmen, H. (1991). Eye specialization in waterstriders: An adaptation to life in a flat world. *Journal of Comparative Physiology A, 169*, 623–632.

Dahmen, H., Wüst, R. M., and Zeil, J. (1997). Extracting egomotion parameters from optic flow: Principal limits for animals and machines. In Srinivasan, M. V., and Venkatesh, S. (Eds.), *From Living Eyes to Seeing Machines*, pp. 174–198. Oxford University Press, Oxford, New York.

Dale, K., and Collett, T. S. (2001). Using artificial evolution and selection to model insect navigation. *Current Biology, 11*, 1305–1316.

Daniilidis, K., Makadia, A., and Bülow, T. (2002). Image processing in catadioptric planes: Spatiotemporal derivatives and optical flow computation. In *Proceedings of the 3rd IEEE Workshop on Omnidirectional Vision (OMNIVIS 2002)*, pp. 3–10.

Douglass, J. K., and Strausfeld, N. J. (2001). Pathways in dipteran insects for early visual motion processing. In Zanker, J. M., and Zeil, J. (Eds.), *Motion Vision - Computational, Neural, and Ecological Constraints*, pp. 67–81. Springer-Verlag, Berlin.

Dror, R. O., O'Carroll, D. C., and Laughlin, S. B. (2001). Accuracy of velocity estimation by Reichardt correlators. *Journal of the Optical Society of America A, 18*(2), 241–252.

Egelhaaf, M., and Borst, A. (1993). Movement detection in arthropods. In Miles, F. A., and Wallman, J. (Eds.), *Visual Motion and its Role in the Stabilization of Gaze*, pp. 53–77. Elsevier, Amsterdam.

Exner, S. (1891). *Die Physiologie der facettirten Augen von Krebsen und Insecten.* Franz Deuticke, Leipzig (Transl. R. Hardie (1989). The physiology of the compound eyes of insects and crustaceans. Springer-Verlag, Berlin).

Franz, M. O., and Krapp, H. G. (2000). Wide-field, motion-sensitive neurons and matched filters for optic flow. *Biological Cybernetics, 83*(3), 185–197.

Franz, M. O., and Mallot, H. A. (2000). Biomimetic robot navigation. *Robotics and Autonomous Systems, 30*, 133–153.

Franz, M. O., Neumann, T. R., Plagge, M., Mallot, H. A., and Zell, A. (1999). Can fly tangential neurons be used to estimate self-motion?. In Willshaw, D., and Murray, A. (Eds.), *Proceedings of the 9th International Conference on Artificial Neural Networks (ICANN'99)*, Vol. CP 470, pp. 994–999. IEE, London.

Franz, M. O., Schölkopf, B., Mallot, H. A., and Bülthoff, H. H. (1998a). Learning view graphs for robot navigation. *Autonomous Robots*, *5*, 111–125.

Franz, M. O., Schölkopf, B., Mallot, H. A., and Bülthoff, H. H. (1998b). Where did I take that snapshot? Scene-based homing by image matching. *Biological Cybernetics*, *79*, 191–202.

Gibson, J. J. (1950). *The Perception of the Visual World*. Houghton Mifflin, Boston, MA.

Gibson, J. J. (1958). Visually controlled locomotion and visual orientation in animals. *British Journal of Psychology*, *49*, 182–194.

Giger, A. D. (1995). *B-EYE: The world through the eyes of a bee (http://cvs.anu.edu.au/andy/beye/beyehome.html)*. Centre for Visual Sciences, Australian National University.

Gluckman, J., and Nayar, S. K. (1998). Ego-motion and omnidirectional cameras. In *Proceedings of the IEEE International Conference on Computer Vision (ICCV'98)*.

Goldsmith, T. H. (1989). Compound eyes and the world of vision research. In Stavenga, D. G., and Hardie, R. C. (Eds.), *Facets of Vision*, pp. 1–14. Springer-Verlag, Berlin.

Götz, K. G. (1964). Optomotorische Untersuchung des visuellen Systems einiger Augenmutanten der Fruchtfliege *Drosophila*. *Kybernetik*, *2*(2), 77–92.

Götz, K. G. (1965). Die optischen Übertragungseigenschaften der Komplexaugen von *Drosophila*. *Kybernetik*, *2*(5), 215–221.

Götz, K. G. (1968). Flight control in *Drosophila* by visual perception of motion. *Kybernetik*, *4*(6), 199–208.

Greene, N. (1986). Environment mapping and other applications of world projections. *IEEE Computer Graphics and Applications*, *6*(11), 21–29.

Greene, N., and Heckbert, P. S. (1986). Creating raster Omnimax images from multiple perspective views using the elliptical weighted average filter. *IEEE Computer Graphics and Applications*, *6*(6), 21–27.

Harrison, R. R., and Koch, C. (1999). A robust analog VLSI motion sensor based on the visual system of the fly. *Autonomous Robots*, *7*(3), 211–224.

Harrison, R. R., and Koch, C. (2000). A silicon implementation of the fly's optomotor control system. *Neural Computation*, *12*, 2291–2304.

Hassenstein, B., and Reichardt, W. (1956). Systemtheoretische Analyse der Zeit-, Reihenfolgen- und Vorzeichenauswertung bei der Bewegungsperzeption des Rüsselkäfers *Chlorophanus*. *Zeitschrift für Naturforschung B*, *11*(9/10), 513–524.

Hausen, K. (1982a). Motion sensitive interneurons in the optomotor system of the fly. I. The horizontal cells: Structure and signals. *Biological Cybernetics*, *45*, 143–156.

Hausen, K. (1982b). Motion sensitive interneurons in the optomotor system of the fly. II. The horizontal cells: Receptive field organization and response characteristics. *Biological Cybernetics*, *46*, 67–79.

Hausen, K. (1984). The lobula-complex of the fly: Structure, function and significance in visual behavior. In Ali, M. A. (Ed.), *Photoreception and Vision in Invertebrates*, pp. 523–559. Plenum Press, New York, London.

Hausen, K. (1993). Decoding of retinal image flow in insects. In Miles, F. A., and Wallman, J. (Eds.), *Visual Motion and its Role in the Stabilization of Gaze*, pp. 203–235. Elsevier, Amsterdam.

Hausen, K., and Egelhaaf, M. (1989). Neural mechanisms of visual course control in insects. In Stavenga, D., and Hardie, R. (Eds.), *Facets of Vision*, pp. 391–424. Springer-Verlag, Berlin.

Heikes, R., and Randall, D. A. (1995). Numerical integration of the shallow-water equations on a twisted icosahedral grid. Part I: Basic design and results of tests. *Monthly Weather Review*, *123*(6), 1862–1880.

Heisenberg, M., and Wolf, R. (1984). *Vision in Drosophila*. Springer-Verlag, Berlin.

Hengstenberg, R. (1982). Common visual response properties of giant vertical cells in the lobula plate of the blowfly *Calliphora*. *Journal of Comparative Physiology A*, *149*, 179–193.

Hengstenberg, R. (1993). Multisensory control in insect oculomotor systems. In Miles, F. A., and Wallman, J. (Eds.), *Visual Motion and its Role in the Stabilization of Gaze*, pp. 285–298. Elsevier, Amsterdam.

Hengstenberg, R., Hausen, K., and Hengstenberg, B. (1982). The number and structure of giant vertical cells (VS) in the lobula plate of the blowfly *Calliphora erythrocephala*. *Journal of Comparative Physiology A, 149*, 163–177.

Hengstenberg, R., Sandemann, D. C., and Hengstenberg, B. (1986). Compensatory head roll in the blowfly *Calliphora* during flight. *Proceedings of the Royal Society of London B, 227*(1249), 455–482.

Hooke, R. (1665). *Micrographia or some physiological descriptions of minute bodies made by magnifying glasses with observations and inquiries thereupon*. Martyn & Allestry, London (reprint by Dover, New York, 1961).

Horridge, G. A. (1992). What can engineers learn from insect vision?. *Philosophical Transactions of the Royal Society of London B, 337*, 271–282.

Howard, J., Blakeslee, B., and Laughlin, S. B. (1987). The intracellular pupil mechanism and the maintenance of photoreceptor signal to noise ratios in the blowfly *Lucilia cuprina*. *Proceedings of the Royal Society of London B, 231*, 415–435.

Huber, S. A., Franz, M. O., and Bülthoff, H. H. (1999). On robots and flies: Modeling the visual orientation behavior of flies. *Robotics and Autonomous Systems, 29*(4), 227–242.

Huber, S. A., Mallot, H. A., and Bülthoff, H. H. (1996). Evolution of the sensorimotor control in an autonomous agent. In Maes, P., Mataric, M. J., Meyer, J. A., Pollack, J., and Wilson, S. W. (Eds.), *From Animals to Animats 4: Proceedings of the Fourth International Conference on Simulation of Adaptive Behavior (SAB'96)*, pp. 449–457 Cambridge, MA. MIT Press Bradford Books.

Indiveri, G., and Douglas, R. (2000). Neuromorphic vision sensors. *Science, 288*, 1189–1190.

Jähne, B. (2001). *Digital Image Processing*. Springer-Verlag, Berlin.

Kelber, A., and Zeil, J. (1990). A robust procedure for visual stabilisation of hovering flight position in guard bees of *Trigona (Tetragonisca) angustula* (Apidae, Meliponinae). *Journal of Comparative Physiology A, 167*, 569–577.

Kelber, A., and Zeil, J. (1997). *Tetragonisca* guard bees interpret expanding and contracting patterns as unintended displacement in space. *Journal of Comparative Physiology A, 181*, 257–265.

Kirschfeld, K. (1967). Die Projektion der optischen Umwelt auf das Raster der Rhabdomere im Komplexauge von *Musca*. *Experimental Brain Research, 3*, 248–270.

Kirschfeld, K. (1972). The visual system of *Musca*: Studies on optics, structure and function. In Wehner, R. (Ed.), *Information Processing in the Visual System of Arthropods*, pp. 61–74. Springer-Verlag, Berlin.

Kirschfeld, K. (1997). Course control and tracking: Orientation through image stabilization. In Lehrer, M. (Ed.), *Orientation and Communication in Arthropods*, pp. 67–93. Birkhäuser Verlag, Basel, Switzerland.

Koenderink, J. J. (1986). Optic flow. *Vision Research*, *26*(1), 161–180.

Koenderink, J. J., and van Doorn, A. J. (1987). Facts on optic flow. *Biological Cybernetics*, *56*, 247–254.

Krapp, H. G. (2000). Neuronal matched filters for optic flow processing in flying insects. In Lappe, M. (Ed.), *Neuronal Processing of Optic Flow: International Review of Neurobiology*, Vol. 44, pp. 93–120. Academic Press, San Diego, CA.

Krapp, H. G., Hengstenberg, B., and Hengstenberg, R. (1998). Dendritic structure and receptive-field organization of optic flow processing in the fly. *Journal of Neurophysiology*, *79*, 1902–1917.

Krapp, H. G., and Hengstenberg, R. (1996). Estimation of self-motion by optic flow processing in single visual interneurons. *Nature*, *384*, 463–466.

Land, M. F. (1989). Variations in the structure and design of compound eyes. In Stavenga, D. G., and Hardie, R. C. (Eds.), *Facets of Vision*, pp. 90–111. Springer-Verlag, Berlin.

Land, M. F. (1997a). The resolution of insect compound eyes. *Israel Journal of Plant Sciences*, *45*(2/3), 79–91.

Land, M. F. (1997b). Visual acuity in insects. *Annual Review of Entomology*, *42*, 147–177.

Laughlin, S. B. (1989). Coding efficiency and design in visual processing. In Stavenga, D. G., and Hardie, R. C. (Eds.), *Facets of Vision*, pp. 213–234. Springer-Verlag, Berlin.

Laughlin, S. B., Howard, J., and Blakeslee, B. (1987). Synaptic limitations to contrast coding in the retina of the blowfly *Calliphora*. *Proceedings of the Royal Society of London B*, *231*, 437–467.

Mallock, A. (1894). Insect sight and the defining power of compound eyes. *Proceedings of the Royal Society of London B*, *55*, 85–90.

Mallot, H. A. (2000). *Computational Vision*. MIT Press, Cambridge, MA.

Mallot, H. A., von Seelen, W., and Giannakopoulos, F. (1990). Neural mapping and space-variant image processing. *Neural Networks*, *3*, 245–263.

McMillan, L., and Bishop, G. (1995). Plenoptic modeling: An image-based rendering system. In *Proceedings of ACM SIGGRAPH 1995*, pp. 39–46.

Müller, J. (1826). *Zur vergleichenden Physiologie des Gesichtssinnes des Menschen und der Tiere*. Cnobloch, Leipzig.

Mura, F., and Franceschini, N. (1994). Visual control of altitude and speed in a flying agent. In Cliff, D., Husbands, P., Meyer, J. A., and Wilson, S. W. (Eds.), *From Animals to Animats 3: Proceedings of the Third International Conference on Simulation of Adaptive Behavior (SAB'94)*, pp. 91–99 Cambridge, MA. MIT Press Bradford Books.

Nachtigall, W. (1968). *Insects in Flight*. McGraw-Hill, New York.

Nayar, S. K., and Boult, T. (1998). Omnidirectional vision systems: 1998 PI Report. In *Proceedings of the DARPA Image Understanding Workshop IUW'98*.

Nayar, S. K., and Karmarkar, A. (2000). 360 x 360 mosaics. In *Proceedings of the IEEE Conference on Computer Vision and Pattern Recognition (CVPR'00)*, pp. 388–395.

Nelson, R. C., and Aloimonos, Y. (1988). Finding motion parameters from spherical motion fields (or the advantage of having eyes in the back of your head). *Biological Cybernetics*, *58*, 261–273.

Neumann, T. R. (2002). Modeling insect compound eyes: Space-variant spherical vision. In Bülthoff, H. H., Lee, S.-W., Poggio, T., and Wallraven, C. (Eds.), *Proceedings of the 2nd International Workshop on Biologically Motivated Computer Vision (BMCV 2002)*, Vol. 2525 of *LNCS*, pp. 360–367. Springer-Verlag, Berlin.

Neumann, T. R., and Bülthoff, H. H. (1999). Minimalistic 3D obstacle avoidance from simulated evolution. In Bülthoff, H. H., Mallot, H. A., and Franz, M. O. (Eds.), *International Workshop on Spatial Cognition: Navigation in Biological and Artificial Systems*, p. 39. Max Planck Institute for Biological Cybernetics, Tübingen, Germany.

Neumann, T. R., and Bülthoff, H. H. (2000). Biologically motivated visual control of attitude and altitude in translatory flight. In Baratoff, G., and Neumann, H. (Eds.), *Proceedings of the 3rd Workshop 'Dynamische Perzeption'*, Vol. 9 of *PAI Proceedings in Artificial Intelligence*, pp. 135–140. Infix Verlag, Berlin.

Neumann, T. R., and Bülthoff, H. H. (2001). Insect inspired visual control of translatory flight. In Kelemen, J., and Sosik, P. (Eds.), *Advances in Artificial Life, Proceedings of ECAL 2001*, Vol. 2159 of *LNCS/LNAI*, pp. 627–636. Springer-Verlag,
Berlin.

Neumann, T. R., and Bülthoff, H. H. (2002). Behavior-oriented vision for biomimetic
flight control. In *Proceedings of the EPSRC/BBSRC International Workshop
Biologically-Inspired Robotics: The Legacy of W. Grey Walter (WGW 2002), 14–16
August, HP Labs Bristol, UK*, pp. 196–203.

Neumann, T. R., Huber, S. A., and Bülthoff, H. H. (1997). Minimalistic approach to
3D obstacle avoidance behavior from simulated evolution. In Gerstner, W., Ger-
mond, A., Hasler, M., and Nicoud, J.-D. (Eds.), *Proceedings of the 7th International Conference on Artificial Neural Networks (ICANN'97)*, Vol. 1327 of *LNCS*,
pp. 715–720. Springer-Verlag, Berlin.

Neumann, T. R., Huber, S. A., and Bülthoff, H. H. (2001). Artificial systems as models
in biological cybernetics. *Behavioral and Brain Sciences*, *24*(6), 1071–1072.

Nilsson, D. E. (1989). Optics and evolution of the compound eye. In Stavenga, D. G.,
and Hardie, R. C. (Eds.), *Facets of Vision*, pp. 30–73. Springer-Verlag, Berlin.

Oppenheim, A. V., and Willsky, A. S. (1983). *Signals and Systems*. Prentice Hall,
Englewood Cliffs, NJ.

Petrowitz, R., Dahmen, H., Egelhaaf, M., and Krapp, H. G. (2000). Arrangement
of optical axes and spatial resolution in the compound eye of the female blowfly
Calliphora. *Journal of Comparative Physiology A*, *186*(7/8), 737–746.

Poggio, T., and Reichardt, W. (1973). Considerations on models of movement detection. *Kybernetik*, *13*(4), 223–227.

Preissl, H., Cruse, H., Luksch, H., Malaka, R., Neumann, T. R., von Sengbusch, G.,
Warzecha, A. K., König, P., Oram, M., Wagner, H., Vollmer, G., Mayer-Kress, G.,
Egelhaaf, M., and Pfeifer, R. (1998). Natural organisms, artificial organisms, and
their brains: The behavior of natural and artificial systems - Solutions to functional
demands. *Zeitschrift für Naturforschung C - A Journal of Biosciences*, *53c*(7/8),
765–769.

Press, W. H., Teukolsky, S. A., Vetterling, W. T., and Flannery, B. P. (1992). *Numerical
recipes in C*. Cambridge University Press, Cambridge, U.K.

Randall, D. A., Ringler, T. D., Heikes, R. P., Jones, P., and Baumgardner, J. (2002). Climate modeling with spherical geodesic grids. *Computing in Science and Engineering*, *4*(5), 32–41.

Reichardt, W. (1969). Movement perception in insects. In Reichardt, W. (Ed.), *Processing of Optical Data by Organisms and Machines: Proceedings of the International School of Physics 'Enrico Fermi', Course XLIII*, pp. 465–493. Academic Press, London.

Reimann, S., Fuster, J. M., Gierer, A., Mayer-Kress, G., Neumann, T. R., Roelfsema, P., Rotter, S., Schoner, G., Stephan, A., Vaadia, E., and Walter, H. (1998). Natural organisms, artificial organisms, and their brains: Emergent properties of natural and artificial systems. *Zeitschrift für Naturforschung C - A Journal of Biosciences*, *53c*(7/8), 770–774.

Riehle, A., and Franceschini, N. (1984). Motion detection in flies: Parametric control over ON-OFF pathways. *Experimental Brain Research*, *54*, 390–394.

Röfer, T. (1997). Controlling a wheelchair with image-based homing. In *AISB Workshop on Spatial Reasoning in Mobile Robots and Animals*, pp. 66–75 Manchester, UK. The Society for the Study of Artificial Intelligence and the Simulation of Behaviour.

Sadourny, R., Arakawa, A., and Mintz, Y. (1968). Integration of the nondivergent barotropic vorticity equation with an icosahedral-hexagonal grid for the sphere. *Monthly Weather Review*, *96*, 351–356.

Schilstra, C., and van Hateren, J. H. (1999). Blowfly flight and optic flow: I. Thorax kinematics and flight dynamics. *Journal of Experimental Biology*, *202*, 1481–1490.

Schölkopf, B., and Smola, A. J. (2002). *Learning with Kernels*. MIT Press, Cambridge, MA.

Schuppe, H., and Hengstenberg, R. (1993). Optical properties of the ocelli of *Calliphora erythrocephala* and their role in the dorsal light response. *Journal of Comparative Physiology A*, *173*(2), 143–149.

Schwind, R. (1989). Size and distance perception in compound eyes. In Stavenga, D. G., and Hardie, R. C. (Eds.), *Facets of Vision*, pp. 425–444. Springer-Verlag, Berlin.

Sherk, T. E. (1978). Development of the compound eyes of dragonflies (Odonata) III. Adult compound eyes. *Journal of Experimental Zoology*, *203*, 61–80.

Srinivasan, M. V. (1993). How insects infer range from visual motion. In Miles, F. A., and Wallman, J. (Eds.), *Visual Motion and its Role in the Stabilization of Gaze*, pp. 139–156. Elsevier, Amsterdam.

Srinivasan, M. V., and Bernard, G. D. (1977). The pursuit response of the housefly and its interaction with the optomotor response. *Journal of Comparative Physiology A, 115*, 101–117.

Srinivasan, M. V., Laughlin, S. B., and Dubs, A. (1982). Predictive coding: A fresh view of inhibition in the retina. *Proceedings of the Royal Society of London B, 216*, 427–459.

Srinivasan, M. V., and Zhang, S. W. (2000). Visual navigation in flying insects. In Lappe, M. (Ed.), *Neuronal Processing of Optic Flow: International Review of Neurobiology*, Vol. 44, pp. 67–92. Academic Press, San Diego, CA.

Srinivasan, M. V., Zhang, S. W., Chahl, J. S., Barth, E., and Venkatesh, S. (2000). How honeybees make grazing landings on flat surfaces. *Biological Cybernetics, 83*(3), 171–183.

Strausfeld, N. J. (1976). *Atlas of an Insect Brain*. Springer-Verlag, Berlin.

Strausfeld, N. J. (1989). Beneath the compound eye: Neuroanatomical analysis and physiological correlates in the study of insect vision. In Stavenga, D. G., and Hardie, R. C. (Eds.), *Facets of Vision*, pp. 317–359. Springer-Verlag, Berlin.

Swaminathan, R., and Nayar, S. K. (2000). Nonmetric calibration of wide-angle lenses and polycameras. *IEEE Transactions on Pattern Analysis and Machine Intelligence, 22*(10), 1172–1178.

Tammero, L. F., and Dickinson, M. H. (2002). The influence of visual landscape on the free flight behavior of the fruit fly *Drosophila melanogaster*. *Journal of Experimental Biology, 205*, 327–343.

Terzopoulos, D., and Rabie, T. F. (1995). Animat vision: Active vision in artificial animals. In *Proceedings of the Fifth IEEE International Conference on Computer Vision (ICCV'95)*, pp. 801–808.

van Hateren, J. H. (1989). Photoreceptor optics, theory and practice. In Stavenga, D. G., and Hardie, R. C. (Eds.), *Facets of Vision*, pp. 74–89. Springer-Verlag, Berlin.

van Hateren, J. H. (1992). Theoretical prediction of the spatiotemporal receptive fields of fly LMCs, and experimental validation. *Journal of Comparative Physiology A*, *171*, 157–170.

van Hateren, J. H. (2001). *Simulations of responses in the first neural layers during a flight (http://hlab.phys.rug.nl/demos/fly_eye_sim/index.html)*. Department of Neurobiophysics, University of Groningen.

van Hateren, J. H., and Schilstra, C. (1999). Blowfly flight and optic flow: II. Head movements during flight. *Journal of Experimental Biology*, *202*, 1491–1500.

von Uexküll, J., and Brock, F. (1927). Atlas zur Bestimmung der Orte in den Sehräumen der Tiere. *Zeitschrift für vergleichende Physiologie*, *5*, 167–178.

Vorobyev, M., Gumbert, A., Kunze, J., Giurfa, M., and Menzel, R. (1997). Flowers through insect eyes. *Israel Journal of Plant Sciences*, *45*(2/3), 93–101.

Walter, W. G. (1950). An imitation of life. *Scientific American, May 1950*, 42–45.

Weber, K., Venkatesh, S., and Srinivasan, M. V. (1997). Insect inspired behaviours for the autonomous control of mobile robots. In Srinivasan, M. V., and Venkatesh, S. (Eds.), *From Living Eyes to Seeing Machines*, pp. 226–248. Oxford University Press, Oxford, New York.

Wehrhahn, C., Poggio, T., and Bülthoff, H. H. (1982). Tracking and chasing in houseflies (*Musca*) - an analysis of 3-D flight trajectories. *Biological Cybernetics*, *45*(2), 123–130.

Williams, L. (1983). Pyramidal parametrics. *Computer Graphics*, *17*(3), 1–11.

Williamson, D. L. (1968). Integration of the barotropic vorticity equation on a spherical geodesic grid. *Tellus*, *20*, 642–653.

Zanker, J. M., Srinivasan, M. V., and Egelhaaf, M. (1999). Speed tuning in elementary motion detectors of the correlation type. *Biological Cybernetics*, *80*, 109–116.

Zeil, J., Nalbach, G., and Nalbach, H. O. (1989). Spatial vision in a flat world: Optical and neural adaptations in arthropods. In Singh, R. N., and Strausfeld, N. J. (Eds.), *Neurobiology of Sensory Systems*, pp. 123–137. Plenum, New York.

Zeil, J., and Wittmann, D. (1989). Visually controlled station-keeping by hovering guard bees of *Trigona (Tetragonisca) angustula* (Apidae, Meliponinae). *Journal of Comparative Physiology A*, *165*, 711–718.

Zettler, F. (1969). Die Abhängigkeit des Übertragungsverhaltens von Frequenz und Adaptationszustand, gemessen am einzelnen Lichtrezeptor von *Calliphora erythrocephala*. *Zeitschrift für vergleichende Physiologie*, *64*, 432–449.

Zollikofer, C. P. E., Wehner, R., and Fukushi, T. (1995). Optical scaling in conspecific *Cataglyphis* ants. *Journal of Experimental Biology*, *198*, 1637–1646.

Titus Neumann studied Computer Science at the University of Stuttgart, and Cognitive and Physiological Psychology at the University of Tübingen, Germany. He received his Diploma in Computer Science from the University of Stuttgart in 1997. His Diploma research was conducted at the Max Planck Institute for Biological Cybernetics in Tübingen, where he focused on algorithms for three-dimensional self-motion detection. From 1997 to 1998 he was a research associate of the DFN (Deutsches Forschungsnetz) working on a virtual reality-based distributed simulation laboratory. His doctoral work at the Max Planck Institute for Biological Cybernetics involved interdisciplinary studies in insect-inspired machine vision, visual flight control, computer graphics and simulations, as well as spherical image acquisition and processing. His research has been partially funded by stipends from the Max Planck Society and the Flughafen Frankfurt Main Stiftung. He is currently a research scientist in the Department of Bioengineering at the California Institute of Technology in Pasadena, California.